HACK YOUR HORMONES

EFFORTLESS WEIGHT LOSS, BETTER FOCUS, DEEPER SLEEP, MORE ENERGY

DAVINIA TAYLOR

WITH A FOREWORD AND CONTRIBUTIONS
BY DR MOHAMMED ENAYAT

First published in Great Britain in 2023 by Orion Spring
an imprint of The Orion Publishing Group Ltd
Carmelite House, 50 Victoria Embankment
London EC4Y 0DZ

An Hachette UK Company

1 3 5 7 9 10 8 6 4 2

A CIP catalogue record for this book is
available from the British Library.

ISBN (Trade Paperback) 978 1 3987 0961 4
ISBN (eBook) 978 1 3987 0962 1
ISBN (Audio) 978 1 3987 0963 8

Typeset by Input Data Services Ltd, Bridgwater, Somerset

Printed in Great Britain by Clays Ltd, Elcograf S.p.A.

www.orionbooks.co.uk

Dedicated to every woman who's ever been told she's 'hormonal'

CONTENTS

FOREWORD: By Dr Mohammed Enayat ix

INTRODUCTION: What We Get Wrong About Hormones 1

Why Can't I Sleep? 13

Why Can't I Stop Eating? 61

Why Does It Feel Like I'm Losing My Mind? 109

Why Do I Feel So Low? 143

Where Is This Rage Coming From? 177

WTF Is Up With My Hormonal Cycle? 205

A Word Before I Go . . . 247

APPENDICES

Glossary 249

Recommended Shopping List 259

Recommended Suppliers 268

FOREWORD

I've been lucky enough to work with Davinia for some time, both as a doctor and a friend. Her approach to changing her biology, and transforming her life, exemplifies what is made possible by being curious, and by testing out simple lifestyle and supplementation changes. *Hack Your Hormones* is a symptom-led guide, sharing her science-based approach in language that is straightforward and easy to understand. If you follow Davinia on social media, you'll know she is honest, frank and bares all. Her honesty helps all of us become more curious about our own biology, and to make positive changes.

As a doctor educated in the UK, I felt that I was primarily trained to identify and treat disease, rather than to focus on optimising the health of my patients. My life and my practice began to change when I stepped outside of the standard expectations of 'disease management' and began to ask myself questions about what true health looks like. Davinia's approach is the same – it's not simply about treating symptoms, but about aiming to feel our very best.

Having established a functional medicine clinic that uses many of the techniques shared in this book I can assure you that hundreds of my patients have discovered how taking control of their health has helped improve their mood, energy, sleep, digestion, immunity, libido and more. Not to mention the confidence they gain from feeling informed and empowered, and from truly understanding how their bodies work.

Functional medicine is a holistic approach to health which works on all systems of the body, including gut health, inflammation, micronutrients and hormones. It uses nutritional and lifestyle interventions to tackle and correct root causes of imbalances. This type of medicine, when delivered well, works alongside traditional medicine, and is not necessarily a replacement for it. If you're interested in learning more, there are a growing number of functional medicine-trained physicians and services that integrate this holistic approach with trained nutritional therapists, integrative psychiatry and psychology.

Making smarter, more informed decisions about your hormonal health requires time and education, and that applies to you and also to your doctor. Our understanding of hormones has evolved so much, and continues to evolve all the time. It was once standard practice for doctors to prescribe the pill to any teenager who was experiencing issues with their periods. These days I would first want to investigate any hormonal imbalances that may be the root cause of menstrual pain or irregularity. And in recent years our knowledge of the benefits of hormone replacement therapy for perimenopausal and menopausal women has come on in leaps and bounds.

Awareness is key, and I believe this book is a tool that will help you develop better awareness about your own hormonal balance, or imbalance. You may need additional support from a medical professional, and throughout this book I will offer some guidance on how to talk to your doctor if you suspect you may have a hormonal imbalance. But in short, if you are worried, please go to see your doctor – there is so much they can do to help.

The sooner you start your journey towards better health, the better and longer-lasting the outcomes. Please don't wait until you're sick.

I'm a firm believer that the patient is the true expert in how they feel. Often the patient has the answers to their health issues themselves – they just need some data points and some direction to get there. *Hack Your Hormones* will help you put the power back in your own hands, and direct you to take steps towards improving how you feel.

You're not just doing it for yourself; a healthier society is a happier, more content and more productive society.

Dr Mohammed Enayat
GP and Functional Medicine Physician
Founder of HUM2N

INTRODUCTION

WHAT WE GET WRONG ABOUT HORMONES

WHAT I GOT WRONG FROM THE VERY START . . .

I started messing with my hormones before I even understood what they were.

I was fourteen when I got put on the pill for a bit of teenage acne. I wanted to clear up my skin and feel a bit more confident, and going on the pill was just what you *did* back in the 1990s. My doctor didn't ask me any probing questions, and there was definitely no talk of potential side effects. But if I'm being honest, going on the pill made me feel really grown up – I didn't think about the fact that it was mucking about with my hormone levels.

Of course, me being me and quite scatty, I didn't even take it properly. I'd forget to take it for days and then throw five days' worth down my neck all at once and hope for the best. Then I'd do that thing, when you're a teenager going on holiday and you just keep on taking it so you don't come on your period. (Not only that, I didn't even know for years that the period you get on your pill isn't even a real period!). So, looking back, I think my hormonal imbalances started off early.

Years later, I went through IVF treatment when I struggled to conceive my first son. It's no secret that I had a very public addiction problem and massive depression after he was born. Now I know this was driven by a severe hormone imbalance after IVF – but it wasn't considered then at all. I mean, *seriously?* Why were my hormones never tested, why was I put on Valium and a twelve-step programme instead? Why did no one look at my oestrogen and progesterone? Why did I feel so frightened that I had to drink just to calm myself down? Why was my cortisol not looked at? I went up and down Harley Street to find the 'best' doctors and instead was put on bipolar medication. I just think now – why? Well, because nobody touched hormones fifteen years ago.

I've seen the worst side of what can happen when your hormones are completely out of whack. I became an addict, had a meltdown, and lost custody of my son. Even fifteen years on I still get legal letters that will throw my mental health diagnosis and past addiction issues at me. There was never any consideration of the root cause, it was just a case of 'you're insane and therefore untrustworthy'. I hit rock bottom, but knowing what I know now, there was no need for me to feel this way. The root to all our behaviours is hormones and that's why they're so important. I don't want anyone else to get to where I did. Hacking my hormones has honestly saved my life.

. . . AND HOW I BEGAN TO PUT IT RIGHT

A few years ago, I'd got through my alcohol addiction, but I was an overweight mother of four and not enjoying my life. I was on a ton of antidepressants, feeling bloated and flatlined. Things first started to change when I watched a BBC documentary about depression. In the programme, they talked about how people can pep up their dopamine levels naturally with cold water exposure. That was a bit of a lightbulb moment for me; I thought, if I can't even be bothered to try a cold shower, what can I be bothered with? So, I did it a few times, and don't get me wrong, I hated it! (I still don't *like* it, it's horrible.) But I did feel a little sense of achievement, and afterwards, I went downstairs and tidied up the kitchen. Cause and effect! I didn't realise it at the time, but I'd started to hack my hormones in a positive way.

Fast forward a couple of years, I realised my PMT was getting worse, so I started tracking my cycle and my mood on a period app, to see if I was perimenopausal. I started digging more and more, reading up about all our hormones and what they control. What I discovered was completely life-changing. I found out that hormones aren't just the demonic derogatory curse that so many people think they are. In actual fact, when our hormones are balanced, they can make us feel fantastic, sleep deeply, giggle, feel cosy and safe, motivated and energised – all the good stuff.

When I understood that balancing your hormones can make you happy, that's when I realised that it had to be the next step for me. I made it my goal to learn and understand everything I could about hormones so I could achieve my goals of being a better mum, wife and friend. I've taken all this knowledge and then biohacked my way to find the right balance: for me, that's a combination of detox, reducing inflammatory foods, running, light exposure, saunas, supplements, breathwork, nootropics, kombucha, intermittent fasting, gut health and, yep, cold exposure (sorry).

I'll go into more detail about these in the book, but what's important is to discover what works for YOU as an individual, and tracking it. It's not about one size fits all, it's not about editing your genes – it's taking a holistic approach to fix your health and live your life optimally.

SO WHAT THE HELL ARE HORMONES?

Simply put, hormones are our body's chemical messengers. They're created by special cells in our endocrine glands and are released into our bloodstream to send a message to another part of our body. We're all familiar with the sex hormones like oestrogen and progesterone that regulate fertility and also affect weight gain. But hormones control so much more than reproduction. They pretty much look after everything! Our stress response is governed by the adrenal gland, which produces adrenaline and cortisol. Our thyroid gland releases hormones that regulate our energy. There are loads of others, including the pineal, thymus and pituitary, and they all have vital roles to play in our health. This probably all sounds a bit too science-y right now, but I'm going to keep it really simple throughout, I promise.

Basically, how our hormones work dictates our mood, our energy levels, how we sleep, our appetite, how patient we are with our family, friends and colleagues, how motivated we are to exercise, and even how we react to a crisis. Hormones are not just about menopause and PMT – it goes way, way deeper than that. They are the building blocks of our physicality AND our personality. And when they're out of whack, they can really affect us, making it hard to sleep, to eat well, to feel good, to concentrate. That's why this book is focused on the symptoms you might be dealing with, and we can dive into what hormones are behind these and what you can do to hack them.

STOP FEELING THE SHAME – IT'S YOUR HORMONES!

If you've tried to fix your problems before now and it hasn't worked, don't blame yourself. There's way too much shame around about things like obesity, depression and addiction. People feel guilty that they can't just 'white-knuckle' their issues, and lose weight, feel better, stop drinking, whatever it is, just by willpower alone. But all these issues are driven by hormones – it's NOT your fault. You cannot beat yourself up for dodgy chemistry in your body.

Like I said, hormones drive everything. They control our behaviours, our urges, our attitude – pretty much anything you can think of that we do and feel as humans has a load of hormones dancing around behind it. For example, if you're eating too many crap foods, your sugar addiction may be driven by an insulin resistance. It's not because you 'lack willpower' – I believe sugar is more addictive than crack cocaine! If your dopamine levels are wrecked, you can find it hard to get out of bed in the morning and lack drive. In my opinion my high dopamine drivers were behind my alcohol addiction, because I was always craving that next high, and by high I mean 'normal' levels to regular people..

So, let's take that futile mantra about so-called willpower and draw a line under it. It's a non-starter. Now's the time to take easily actionable protocols, and what we're going to do in this book is figure out what's going wrong with your chemistry, fix it and then you can be a super human, whatever that means to you. We have a kitchen cupboard full of these incredible ingredients and we can cherry pick which hormones we want to hack in order to live our best life.

BUSTING THE MYTHS ABOUT HORMONES

One thing that really annoys me is all the stigma and bollocks that's attached to hormones, and what they do. Firstly, there's that old myth that hormones are just all about the negative effects. We're all used to people (let's be honest, women) being described as 'hormonal' as a criticism. And yes, fine, our hormones can make us moody or stressed out, but they can also do the opposite: they can make us feel energised, safe, they can make us burn fat, they can clear our skin up, they can give us great hair.

Unfortunately, the word hormones gets thrown around without people understanding what the hell they're on about. I mean, unless you've seen my bloodwork and dry urine test, what hormones are you talking about if you say I'm 'hormonal'? Is it my cortisol or am I insulin resistant – let's be specific! Or, maybe, here's a thought, just don't use it as an insult at all.

Another myth is that hormones are only something to worry about if you're female, going through puberty or menopause. This just isn't true – we ALL have fluctuating hormone levels. All of us, men and woman, every single day have cycles of hormones doing a dance with each other, interacting and causing effects constantly. Balancing and understanding this symphony of hormones is important for every single one of us.

NO SUCH THING AS 'NORMAL'

One thing I really want to say, too, is that this book isn't about all of us achieving some sort of unified, 'perfect' hormonal state. We're all wired completely differently and what works for some of us won't work for others. I've got lots of friends whose ideal state would be to curl up with a book or watch TV with a cat on their lap. Lovely, but that's completely NOT my comfort zone! I don't sit well with boredom or relaxing, because I like being in a hyper, really dopamine-driven state. So doing yoga doesn't work for me, but running does. And that's fine, that's my hormonal makeup.

I like to think of all of us humans as part of a tribe, where every member brings something different to play. We fill in the gaps for each other. I mean, thank God I didn't have quiet, creative kids, because I'm crap at crafts and I'm shit at baking, but bloody hell I can take my three sons for a run and to an obstacle course! We have different hormonal makeups and different strengths, so let's work to our gifts and not try to emulate somebody else.

Another thing to say is that testing our hormone levels can be really useful but comparing them to others isn't necessarily vital. For example, men's testosterone levels that were considered chronically low twenty years ago are now 'normal', so how is that helpful? I mean, *WTF?* And hormone levels differ globally – what's normal for someone in France is different to Singapore is different to New York is different to Wigan! I think we should be less focused on numbers and more focused on symptoms, which is why this book is symptom-led. Let's get down to reality – are you anxious? Are you sleeping badly? Are your eating habits detrimental? Right, now we've got something to work with.

WHAT THIS BOOK WILL DO FOR YOU

Hacking my hormones has kept me sober, kept me lean, made me really productive and just more confident as a woman. I'm still scatty and have ADHD, but that's fine! It's about progress, not perfection. You'll learn about ADHD later – yes, it's hormone driven too!

I want this book to be a manual for you to figure out what's behind your issues and show you the simple changes you can make to help yourself. You might notice that a couple of chapters – the sleeping and eating ones – are a little longer than others, but that's because they're HUGE topics, with a lot going on hormone-wise! I've tried my best to include everything that I've read and researched about these issues, and make it as easy as possible to understand and action for you and your life.

My approach is not about dieting, or taking things away from your lifestyle; instead, it's about adding positive habits. I hate phrases like 'anti-anxiety' – let's think of it instead as 'pro-safety'.

Of course, we're going to go over the basics when it comes to hormones, but we're also going to go to a deeper level so you have the vocabulary to express yourself. I want to arm you with the information you need so you know how to understand the root causes, target your symptoms and feel confident speaking to your GP if you need to. We've been screaming underwater for so long – women have been called insane, been sectioned, been given electric shock therapy and killed themselves in the past because of hormones that nobody understood yet. But we've come so far, and now we understand that when we're feeling 'hormonal', it's because we're out of balance.

This book will show you how our hormones are an amazing tool at our fingertips, as long as we hack them right.

THE BRILLIANT DOCTOR E

Throughout this book, you'll also read contributions from Dr Mohammed Enayat, otherwise known as Dr E. He's an amazing GP, with extensive training in functional and regenerative medicine. He's now a specialist in preventative medicine, where the focus is on achieving vitality with our health, not just getting rid of disease. I first came across Dr E when I was looking into hyperbaric oxygen chambers a few years ago. I'm always looking for ways to extend my dad's life and improve his health, so I started digging online and came across Dr E's clinic and his ethos seemed to match mine – he wasn't too woo-woo!

Since then, we've developed a great working relationship as we're both really into biohacking and what it can do for our health. 'Biohacking' might sound scary and overly technical, but it's really about getting our body into a state of homeostasis (balance), through a process of actionable protocols and techniques to find out what works best for us as individuals to optimise our personal health. I love chatting to Dr E about what's coming next, as I'm always on the lookout for the latest thing, and he's so ahead of the game, with a team of incredible professionals.

Dr E has approved all the medical info and advice in this book, but he's also going to give you an insight into what you can ask the NHS for, and give you the vocabulary and ideas to work with your GP. He'll also suggest what you can action outside the NHS. I love the NHS as much as anybody, but it's a massive machine and is set up to treat the ill. Our goal is slightly different – instead of treating illness, we want to be really *well*!

BALANCING THE EVOLUTIONARY MISMATCH

Our crazy modern lives have moved on far too fast for our hormones to catch up. Back in the day, thousands of years ago, we'd have had our kids at sixteen because that's when we're most fertile, and by the time we were forty, we'd be the wise old woman of the village, NOT dealing with screaming little kids or work stresses! But that's not what we do now, is it? The combination of stress, lifestyle and how society has evolved makes things difficult. So, we need to hack our hormones because we're living in an evolutionary mismatch!

Trust me, we can learn how to survive our prehistoric hormones in a modern-day world. Hallelujah, we are finally starting to understand them now, and we're thousands of years old as a species! Whatever age we are, our hormones are affecting us and luckily we now have the science and tech to live longer, so let's live longer better.

In this book I'm going to give you some insights and ideas as to what might be going wrong and some hacks, protocols and information to help you put it right.

PUTTING THE POWER BACK IN YOUR HANDS

This book isn't about turning you into an overachieving supermum, or a CEO. It's about giving you a game-changing confidence increase, where you can feel productive, safe and happy. I want to give you permission to get more out of your life than just 'plodding'. Don't feel shame or guilt if you're not dealing with some huge trauma – our bodies have the same hormonal reaction whether we're being beaten up or have simply lost an important email. We all have different triggers and might not know the reason why, but that doesn't matter.

Ultimately, it's about taking control of your body and putting the power back in your hands so you can control your mood, sleep and life. And if you have a bad night, or feel like shit, or end up eating a load of crap one day, don't worry – we're all human, we're not perfect. I have a tricky hormonal system that gets out of whack and I need to rein it in sometimes. I can easily get overwhelmed by stress and changes to my routine – especially when I'm away from home. It's ironic, but going on holiday can really throw me out of whack and make me feel hyper-hunted and low. I've definitely found that the older I get, the more susceptible I am to the knock-on effects of LIFE. The difference is that nowadays, I have the knowledge in my artillery to help myself get back to a state of balance, rather than reaching for a bottle of wine, a slice of cake, or both.

So, let's get our hormones into the optimum place they can be for us. Because if *I* can feel better – and this is coming from someone who was extremely physically and mentally ill – then anyone can. This is proper, real empowerment, not tampon-ad empowerment! This is *your* body, nobody can look after it like you can.

WHY CAN'T I SLEEP?

IN THIS CHAPTER, WE'LL FOCUS ON THESE HORMONES:

Serotonin: *makes us feel cosy, safe and content*
Melatonin: *the master hormone behind sleepiness*
Cortisol: *regulates our stress response*
Adrenaline: *our fight-or-flight energy hormone;* ***noradrenaline/ norepinephrine:*** *adrenaline in the brain that helps us wake up*
GABA: *stops anxiety and helps us relax*
Vitamin D: *actually a hormone that helps regulate our sleep*

We're kicking off here because getting great sleep is the number one thing we can do for our health. I definitely didn't prioritise sleep when I was younger. I mean, I used to stay up all night and do weekenders *all the time.* And even if it wasn't that full-on, I'd be drinking a bottle of wine a night, so the quality of my sleep was crap. Lack of sleep also meant I didn't have the energy to tackle the next day. Instead, my overriding emotion was, 'Oh my GOD I'm exhausted, I'm going to have to patch over the day with coffee and croissants.'

I think I was also absorbing the general attitude to sleep back then, which was very dismissive. There was this whole 'sleep is for wimps' and 'I'll sleep when I'm dead' mentality, which fed into my approach, which was *who cares?* I honestly still think I've got the after-effects of this sleep deprivation now, as I have low cortisol (get up and go) levels during the day, and if I'm not careful, can get cortisol spikes at night just when I don't want them (I'll explain what cortisol does in a bit).

WHAT RUBBISH SLEEP DOES TO ME

I always feel *so* crap when I sleep badly. It's a form of torture when you can't sleep, or your sleep is interrupted (and it actually *is* used as a form of torture – sleep deprivation is how prisoners of war are broken down!). When I was pregnant, I had insomnia and, of course, I was forever getting up to wee in the night. But that was nothing compared with what it felt like when I first had babies and was dealing with those broken nights. It came as a complete shock how horrific it felt.

I'd get a horrible taste in my throat, have massive brain fog and be *very* short tempered. I'd have no patience at all, and I think if you're pretty feisty like I am, that's the first thing to go when you're sleep deprived! My mum was completely different to me, because even though she had terrible sleep issues she managed to stay calm. I'm a bit more like my dad; if he doesn't sleep, he blows his top. It's that awful feeling of being out of control – I go into complete panic mode if I haven't had a good night's rest.

These days, thankfully, I sleep really well, but I still have bad nights every now and then. I'll get insomnia sometimes when I'm in the early days of my menstrual cycle, and when I'm out of my routine. We humans are absolutely creatures of habit, so when I'm travelling, I don't sleep as well as I do at home. Even though it's a complete mess half the time, home is my safety net. When you sleep in a hotel, it puts you on high alert, because you're not used to the environment and the noises.

Not long ago I was staying overnight in London and I thought it would be amazing to have a quiet night without the kids, but it *really* messed with my sleep. There were all sorts of lights coming through the doors from the corridor, weird noises from the TV, air-con and the street outside; I was lying there with my eyemask on fretting like a nana! It was far from glamorous jet-setting fun – there really is no place like home.

SO MANY OF US ARE STRUGGLING

Sleep problems can take on different forms, whether that's insomnia, waking up in the night (or too early in the morning), unrefreshing sleep or bloody kids stopping you from getting a good night's rest. I know loads of my friends who struggle with sleep, and some of them rely on alcohol to nod off – not in the sense of getting blind drunk, but having a few glasses of wine to lower their anxiety. It can just end up exacerbating the problems though, with worse quality sleep and then you wake up in the night when it wears off. I know how tough it is though, so there's no judgement.

On my Instagram I hear from loads of people who are dealing with sleep issues, too, especially that racing mind that stops you falling asleep or popping awake in the middle of the night and then not being able to nod off again. It really is an absolute crisis for so many of us and it's such a lonely place to be at 2 a.m., freaking out about how you'll cope the next day. I'm an arsehole without sleep, I hate my life, I have no confidence, I have no productivity and I just want to give up on everything. It really is the worst thing for me. I'd rather be hungry than tired – and that's *really* saying something.

WHY HACKING THE SCIENCE WILL HELP YOU SLEEP BETTER

More than two thirds of the UK population suffer from disrupted sleep and nearly a third are dealing with insomnia, so if you're struggling, YOU ARE NOT ALONE! It's an epidemic, but we can hack it. So, if this is you, do not worry – we are now able to use science to help you. It's not just a case of going to your doctor and getting sleeping pills, which make you feel worse and more hung-over the next day (not to mention how addictive they can be). It's just knowing a bit of the science so we can understand WHY it's happening, and a massive, massive part of this is our hormonal balance.

What we're going to explore in this chapter is how our hormones affect our sleep, and how you can hack them to help yourself. The secret to good sleep is not about mindfulness and crystals – we're talking about real science, real research, real results. To be honest, I find all this 'be kind to yourself' stuff really unhelpful – I'm *way* more cynical than that. I need to know the science, understand the proof and not just think I'm wasting my fucking time sitting around chanting! And if you're trying the woo-woo stuff and it's not working, you can end up beating yourself up even more.

I'm here to say it's OK, it's not your fault, if you can't sleep, it's highly likely your hormones are out of whack. Your issues are probably to do with the light and noise pollution, what you're eating and your body clock. We're living in a day and age where our lifestyles are completely geared against us. If you're lying in bed with that feeling of racing panic at 1 a.m., that's adrenaline, and maybe cortisol too, which are both a completely natural part of our body's chemistry – they're just not meant to be active at that time. We're going to get your body's chemistry back into balance because ultimately your body wants you to survive and thrive.

WHY SLEEP MATTERS

Sleep literally underpins everything to do with our health. Our bodies need sleep to repair muscles, consolidate memories and restore balance to the body, like bringing our insulin (which regulates blood sugar) into alignment. We go into detox mode when we're in deep sleep, flushing out waste and toxins from the brain, releasing hormones, reducing inflammation and allowing the body to heal and repair in our 'rest and digest' state. Sleep is absolutely essential for pretty much everything.

When we're not sleeping well, we know how bad it makes us feel the next day – irritable, weepy, fatigued, angry, no energy, all that horror. But also, lack of sleep can have longer-lasting effects on us, too, like weight gain, depression, and even cardiovascular disease. It's no wonder restorative sleep is cited more and more as paramount to achieving good physical and mental health.

Thankfully, the overall attitude towards sleep has massively changed in recent years. As well as the medical understanding, on the social front, people are talking about great sleep more and more. Look, you've even got Kim Kardashian and Gwyneth Paltrow comparing their sleep scores from their Ōura rings on Instagram now! You'd have never had that a few years ago. We're beginning to revere sleep now, it's like it's the new rock'n'roll.

THE DIFFERENT SLEEP STAGES AND WHAT THEY DO

When we sleep, we go through different stages and repeat these during the night, because each stage performs different functions in our brains and bodies. I used to think a graph tracking our sleep would look like an upside-down bell curve, where we 'sink' down into sleep and then rise back up gradually through the night to waking up. But actually, graphs tracking our brain activity show that it goes up and down throughout the night as we work through the different sleep stages, releasing different hormones and different brain waves. Like Kim K and Gwyneth, I've got an Ōura ring and it's completely fascinating to look at my sleep wave cycles in the morning.

Our sleep stages go like this:

Stage 1

This is the transitional phase where you fall asleep. You're in almost a meditative state as your brain releases beta waves and you might experience that weird falling feeling as your body moves into a sleep state. At this point, you're still mostly alert, and might not even feel as if you're asleep at all.

Stage 2

The light sleep stage, which lasts twenty to thirty minutes – here your brain makes periodic surges in activity called 'sleep spindles', which is why dreaming is possible now. You're in a true sleep, with your body temperature dropping and your heart rate slowing. You can still be woken up easily in this stage though.

Stages 3 and 4

You're deeply asleep here, and your breathing becomes very stable. (This is where I get annoyed by Matthew and his breathing – the other night I had to get into my son Jude's bed instead because I was so pissed off Matthew had dropped into Stage 3 sleep before me!) Your muscles are completely relaxed, your blood pressure and body temperature decreases, and your brain releases slow delta waves. Your body uses this time to repair damaged tissue and release hormones. It's very hard to wake up from this stage, which generally lasts thirty to forty minutes.

REM (Rapid Eye Movement) sleep

This is the biggie, the REM sleep that's crucial for regenerating your brain's nerve cells. Your brain is awake but the rest of your body is asleep – your muscles from the neck down are actually paralysed here to prevent sleepwalking. Your eyes will flicker rapidly under your lids, as this is where you do most of your dreaming. This stage lasts only about thirty to forty minutes and once it's finished, we pop right back to Stage 1 sleep.

You cycle through all these stages four to five times throughout the night – *if* you're getting a good night's sleep, that is!

SO WHERE DO HORMONES COME IN?

Hormones regulate our sleep, and if they're working well, we sleep well. If they're not, then we've got a problem. There are so many hormones that influence our sleep health, but I'm just going to focus on a few of the most important ones.

Serotonin

A really good place to start is with serotonin, which is produced by the pineal gland located deep in the brain. It's know as the 'happy hormone' as it regulates your mood. Serotonin makes you feel safe and calm, cosy and homey – the type of feeling where you don't want to run out to a nightclub, you're happy where you are watching *Coronation Street* with a cup of tea.

A fascinating fact which will be a biggie throughout this entire book is that we make serotonin from the nutrients we eat – literally 95 per cent of serotonin is manufactured in the gut.*

Serotonin starts off as tryptophan, an essential amino acid that we need for growth when we're kids, and a whole load of metabolic functions that affect our mood and behaviour as adults. We can't produce tryptophan ourselves, so we get it from tryptophan-rich foods. If you're having trouble sleeping I recommend upping your intake of these foods to help the release of serotonin. Some great sources include:

- turkey and chicken
- grass-fed red meats
- tuna
- cheese, milk and other full fat dairy

* https://www.apa.org/monitor/2012/09/gut-feeling

- bananas
- sour cherries
- bovine collagen peptides

When we eat these foods, our bodies convert the tryptophan into 5-HTP, a different chemical. This then turns into serotonin, which then a few hours later, turns into the most important hormone for sleep . . . drum roll please . . .

Melatonin

This is the big one, the master hormone behind sleepiness. Our body releases melatonin, which starts to happen naturally about two hours before sleep. It's the final phase of that tryptophan–5-HTP–serotonin–melatonin sequence, and we aren't going to fall asleep until we get it. (I found out recently that babies don't produce melatonin until they're three months old, which is why when they're newborns they want to party all night and destroy you.)

You need to have that cosy, safe, all-is-well serotonin for your body to trip that into melatonin and let you fall asleep. If you're filled with anxious thoughts or up late on your phone with all the lights blaring, you're going to stop that melatonin from doing its trick.

This is because light levels have a massive impact on melatonin. The pineal gland (yep, that one again) increases production of melatonin in the evening as it gets darker, and peaks in the middle of the night. It falls to its normal daytime low by early morning. This is why exposure to the right sort of light at the right time of day can really help regulate your melatonin levels, and I'll show you how to do this later in this section.

(Another note on melatonin – you can't get it as a supplement very easily in the UK, which I think is crazy, as in the States it's a really popular over-the-counter remedy for sleep issues and jetlag. Here, you have to get a prescription from your doctor for melatonin. I mean, I can buy vodka in a BP garage on the M6, but I can't be trusted to buy a safe sleep hormone from the pharmacy? Bonkers. I personally prefer melatonin drops to pills.)

Adrenaline

Adrenaline is our 'fight-or-flight' hormone, regulated by the adrenal glands which are just above each of our kidneys. It's released when our brain sends a message to these glands that a stressful situation is happening. Imagine that nervous energy you get just before a race or an exam, where you experience a racing heart, tense muscles and maybe even a bit of sweat? That's adrenaline.

Adrenaline can be really helpful as it's designed to keep us safe. It gives us a surge of energy, which is essential if we're running away from a massive threat or need some get up and go to get stuff done or hit a deadline. It's not so essential when you're trying to get to sleep – in fact, it's a pain in the arse.

You might also hear about ***noradrenaline***, which can also be called norepinephrine. It's essentially adrenaline that's released in the brain, getting you ready for action by constricting your blood vessels. It supports the fight-or-flight response by increasing your blood pressure, breaking down fat and increasing glucose levels. It also helps maintain your sleep–wake cycles, helping you get up in the morning and sharpen your focus during the day.

Cortisol

Known as the 'stress hormone', cortisol is part of the same family of hormones as adrenaline as it's also produced by the adrenal glands. You don't feel the effects of cortisol as quickly as you do adrenaline because there are a couple more steps involved in producing it. Bear with me, I'm going to get a bit science-y here.

Firstly, the part of your brain called the amygdala has to recognise the stress and send a message to another part of your brain called the hypothalamus. That then releases a hormone to the pituitary gland which sends another message to the adrenal glands to produce cortisol. Flipping heck.

Our bodies will release cortisol once every twenty-four hours in a big cortisol 'pulse' around thirty minutes after we get up. Its main role is to promote alertness when your body goes into fight-or-flight mode and to maintain your sleep–wake cycle so you get the right amount of sleep. Although it often gets demonised, because it's so strongly associated with stress, cortisol is absolutely essential, and your body will continue producing it until the day you die! It gets you out of bed, it gets you sorting the kids' packed lunch boxes and them into the car so they're not late for school – it's a great motivating hormone. I'd be completely exhausted without my natural burst of cortisol in the morning.

Also, cortisol is brilliant at supporting your immune system to fight viruses. Low cortisol equals low energy and immunity. To help with getting up try licorice root, rhodiola and also electrolytes in water.

But! Of course, if your cortisol is out of whack, then you're going to face some problems. If your body produces too much cortisol over prolonged periods at the wrong time, not only will this make you feel on edge, it will make it impossible to drift off to sleep when you need to.

GABA

Or, to give it its full name, gamma-aminobutyric acid. Actually, let's just stick with GABA, shall we? This is an anxiety-inhibiting neurotransmitter (chemical messenger) in the brain that works to block our central nervous system. (Our nervous system is our body's 'command centre' that controls everything from movements and thoughts to automatic processes like digestion and breathing.) GABA slows down our whirring brains and dampens down anxiety, giving us that calm, relaxed feeling that is essential for sleep.

Our brains naturally release GABA at the end of the day, but it's very common to have a GABA deficiency if you're depressed or anxious about sleep. That's a disaster for you, because then instead of feeling mentally safe and calm at bedtime, you're stressing out making mental lists about birthday parties and buying people bath bombs, or whatever it is.

(Again, like with melatonin, you can't buy GABA directly in supplement form in the UK. But you can increase its levels with exercise and eating certain foods, which I'll talk about later.)

A quick note on vitamin D

Now, before you say anything, vitamin D *is* a hormone! Why they call it a vitamin I just don't know (and no wonder there is so much misunderstanding about hormones, if they're calling one a frigging vitamin!). It's produced in the skin, in response to sunlight and the sunlight's interaction with cholesterol under the skin (yes, cholesterol is good – more on that later), and helps absorb calcium from the gut into the bloodstream but is now becoming more known for its role in sleep regulation.*

Lots of studies have linked low levels of vitamin D to a higher risk of sleep disturbances, so boosting your levels of vitamin (HORMONE!) D is really important. However, it also has a suppressive effect on melatonin, which is why if you take it as a supplement, you should take it in the morning, not in the evening when you want to encourage melatonin production.

HOW THESE HORMONES WORK TOGETHER ON OUR SLEEP

Like I've said before, hormones don't work in isolation – they are constantly doing a dance together and having an effect on each other. We need all these hormones; cortisol and adrenaline, serotonin and melatonin, but if they're released at the wrong time of day, they'll stop us sleeping!

Let me show you what the release of hormones would look like over the course of a normal day, assuming you're in a good sleep pattern:

7.30 a.m. Melatonin secretion is at its lowest level, cortisol takes over and you wake up. Your cortisol levels are at their highest right now, motivating you to get out of bed.

12 p.m. Your cortisol levels have dropped off a lot by now, and by mid-afternoon you may begin to feel tired.

* https://pubmed.ncbi.nlm.nih.gov/32156230

6 p.m. Serotonin levels start to rise as you have lovely, calm, polite chats with your kids and partner with candlelight flickering (yeah, right!).

9 p.m. High levels of serotonin mean your body starts to release melatonin, inducing that feeling of sleepiness a couple of hours before bed.

11 p.m. Rising melatonin does its trick and you nod off.

3 a.m. Melatonin levels peak right about now.

HOW MODERN LIFE MESSES WITH US

Now, wouldn't it be lovely if we all lived like this? Everything in harmony and our hormones being released just when they should. Marvellous. But as we know, we don't. Our modern-day lives are playing *havoc* with our sleep hormone levels.

Think about what we put ourselves through on an average day. We get stuck in traffic on our way to work, loads of emails pile up in our inbox, and we don't get enough natural light stuck in an office. Later on, some dickhead cuts us up in traffic, and then we're late picking up the kids, who then moan on about dinner and drive us crazy leaving a mess around the house. While we're trying to sort this, an arsey email from our boss arrives at 7 p.m., sending us into a panic. All of these things trigger adrenaline and cortisol release throughout the day.

This sort of lifestyle means we end up with chronically elevated levels of stress hormones ticking over in us *all the time*. To cope with this, we'll then spend our evenings trying to wind down by scrolling on our phones, or binge-watching Netflix, but the massive exposure to blue light from screens tricks our bodies into thinking it's daytime, and therefore messes with our melatonin release. Bingo, we can't sleep!

Once we're out of our rhythm, it can become a vicious cycle. If you're struggling with sleep, you become more anxious about it. Then, if you're feeling like this, you'll put your body under more never-ending stressors. This will mean you'll have elevated cortisol levels at night-time, which will stop you being able to relax, and instead you'll have that horrible 'tired and wired' feeling. (Some people also report getting adrenaline rushes as they actually try to go to sleep, because your poor brain now perceives going to sleep as a threat.) So, everything is out of whack, you're now dealing with adrenal dysregulation and, of course, you CAN'T SLEEP! Argggh.

It can be a bloody nightmare, I know. But we can hack it. It's all about getting our hormones to release at the right time and working with our natural circadian rhythms – otherwise known as our body clock.

WHY SLEEP IS RULED BY OUR BODY CLOCK

No matter who we are, and wherever we live on the planet, we all have an internal body clock. Every single one of our cells works on a roughly 24-hour clock which dictates what it does, and when. This is how we've evolved over thousands of years; even before humans had invented watches, our cells learned to tell the time before we could! So, we're fundamentally wired to work with our internal body clock, which is completely in tune with the cycles of light and dark.

Think about it. How do cockerels know the right time to start screeching 'cock-a-doodle-do' at dawn every morning? They don't have an Apple Watch alarm! They know because of the light. And this is how we've evolved too, with light and dark levels triggering different hormone releases at different times of the day. These triggered hormones dictate everything, whether we're hungry, energised, anxious, relaxed or sleepy.

Of course, as we've seen, our non-stop crazy modern lifestyles have messed with this and it's having detrimental effects on our sleep. Maybe in 3,000 years we'll evolve again to adapt to this, but that's too late for us! Instead, we need to tune ourselves back into our correct circadian rhythm and get the light–dark cycles in line with our sleep–wake cycles.

THE FOUR THINGS WE CAN HACK TO CONTROL OUR SLEEP HORMONES

You have to hack your body clock by timing when your body releases the right hormones. We want them all to work their magic: cortisol, adrenaline, serotonin and melatonin, but they can only help us if they're in tune with our natural circadian rhythm. There are four really important elements to this in our control, which I'm going to explore in more depth, but essentially, they are:

1. **Light**
2. **Food**
3. **Temperature**
4. **Breathing**

1. LIGHT

WHY GOOD SLEEP STARTS FIRST THING IN THE MORNING

I literally cannot overestimate just how important light is for sorting out your sleep issues. Everything really comes back to this and it's the number one thing you can start to hack easily and best of all FOR FREE! These little things called your eyes are receiving constant signals from the light you expose them to, as our retinas are the gateway to our hormone receptors. What they're exposed to affects the messages they send to our brain, telling it to release this or that hormone.

So, we can hack into our brain chemistry by exposing our eyes to, or avoiding, certain light waves at different times of day. We need to be more in tune with how all these different light waves affect us, because this is simply how we're wired. There are hundreds of scientific studies around on this now, showing the impact of light exposure on our brains and endocrine system. It's nothing to do with being a hippy and getting more in tune with nature (man), it's actually about science showing us that light is a really powerful force and one we've ignored for years. But we're not going to any longer.

GET MORNING LIGHT FOR CORTISOL

Daylight triggers our cortisol release first thing in the morning – fact. Cortisol gives us energy and drive for the day, so we want it when we get up, NOT when we're going to bed. In order to make this happen, the first thing we can do to help balance our cortisol is get outside in the morning and get some natural light. It's really not the same just switching on all the lights in your bedroom, because it doesn't have anything like the same effect.

Light intensity is measured in units called lux, and it's crazy how different light sources have varying levels. Here's how they compare:

- Bright sunny day: 100,000 lux, takes two to three minutes to trigger cortisol release
- Cloudy day (hello, UK): 10,000 lux, so takes thirty minutes to release cortisol
- Bright indoor light: 1,000 lux, takes hours to release cortisol
- Low indoor light: 100 lux – you get the picture.

Windows get in the way, too. I recently read that if you were in an aeroplane, you'd have to stare out of the window for six hours to get the same effect from the light that just being outside for a few minutes would do.

So, it's a matter of triggering your cortisol release at the right time, by using light exposure. Get outside with your coffee first thing and look at the sky (not directly at the sun, obviously, that would be stupid). You don't have to be meditating and omm-ing or saluting the sun, this is about using neuroscience to schedule your hormone release. And I know most of you reading this live in the UK, where it's cold and wet a lot of the time, but as far as I'm concerned, it's never the wrong weather, it's the wrong clothes. I've got a DryRobe, so if it's raining I put that on over my pyjamas when I go outside, which keeps me nice and cosy.

CONTROL YOUR MELATONIN WITH DARKNESS

It might be tempting when it's really sunny, but I recommend you don't wear sunglasses in the morning. Wearing them will trigger melatonin release when you don't want it, because they trick your eyes into thinking it's getting dark, and it's time for bed. I know it can be hard at first if you're used to wearing sunglasses every time you go outside, but just spending a few minutes without them in the morning can make SO much difference to your hormones.

Instead, you want to trigger your serotonin–melatonin pattern in the evening, and that really starts with when the natural light goes down. You should start dipping your light levels in line with the seasons, so earlier on in the winter, and later in the summer. Dim the lights if you can, turn off glaring overhead lights and use warmer, softer, side lamps to mimic the natural evening light. You can even use candles to create a darker ambience, though not those stupid expensive ones that cost a fortune and are full of perfume and petroleum which can disrupt your hormones, just use natural soy or beeswax unscented ones and burn organic essential oils if you want a fragrance.

So, dim the lights as the day goes on. Put your sunglasses on in the evening – it might feel weird but stick with it for about a week to reset. At night-time, make your bedroom as dark as it possibly can be – blackout blinds are amazing for this. Unbelievably, light on your skin can stimulate cortisol even if your eyes are closed, so you've got to get full blackouts rather than thinner curtains or blinds.

Any kind of light at night is going to interfere with your

melatonin levels. So if you've got a telly in your room, put a sticker or electrical tape over the red standby light (or better still, turn it off at the mains). Even when you're going to the toilet in the night, try and keep the light off (and aim carefully). You want to do everything you can to help your body produce melatonin, that lovely sleepy feeling, at the right time.

NATURAL LIGHT VS BLUE LIGHT

Natural light is referred to as 'full spectrum', which means it contains all the colours of the rainbow, and even some light waves we can't see, like ultraviolet and infrared. Our screens – phones, tablets, laptops – are all emitting synthetic, blue-spectrum light, which mimics natural daylight and our prehistoric brains can't tell the difference. So instead of your brain thinking, 'Oh right, time to release some melatonin and have a lovely sleep,' it's had three million years of evolution telling it, 'Right, here's some daylight, time to go out and hunt!'

So, if we're scrolling on our phones or watching Netflix at 10 p.m. (which look, OK, I still do – it's my happy place, sorry), we're confusing our brains. This is a *disaster* for our hormone release, because it stops us secreting melatonin just when we should be. I might still fall asleep after a Netflix binge, but even then it won't be the best restorative sleep, as I'll have some residual cortisol racing around my system.

It's amazing to see the effect of restricting blue light in action. Matthew and I went camping a few years ago and my little night owl was asleep by 10 p.m. because he had no blue light. (It was annoying, though, because he'd fallen asleep before me in the tent!) He dropped off as quickly as if he'd had a sleeping pill, whereas normally he doesn't go down until two in the morning (he's a complete and utter vampire). But because we had no phones or tablets, they weren't messing with his melatonin release, and he had an incredible night's sleep.

Look, I'm not a monster, I don't expect you to ban all screens from 6 p.m. onwards and sit there in the dim light crocheting or whatever. Let's face it, we all love our binge TV, and I've definitely still got screens blaring in the evening! But if you're struggling with sleep, and still want to use screens, you can hack it by wearing blue-light-blocking glasses – for me, they're a complete game changer. You can buy them online now for as little as a tenner, and they work by stopping the blue light waves from our screens that trick our brains into thinking it's daytime. I wear them in the evening when watching telly, or on my phone – I promise, there are some nice ones around now that won't make you look like a cyborg!

2. FOOD

HOW WHAT AND WHEN YOU EAT IMPACTS YOUR SLEEP

Another really powerful hack for improving your sleep, is what you eat – and just as importantly, WHEN you eat it. Like I mentioned before, you want to support your levels of serotonin, the happy hormone, which then becomes melatonin, the sleep hormone – and you want that to happen at the right time of day, i.e. BEDTIME! It's amazing but true: we can protect our sleep with what we eat.

If you've got trouble with going to sleep, it's likely you've got low serotonin levels. Previously, there was some uncertainty in the scientific communities on how important healthy levels of serotonin were for good sleep, but it's now been proved beyond a shadow of a doubt. Researchers at CalTech in the States found that serotonin in the brain has a direct link with building up 'homeostatic sleep pressure',* which is your natural tiredness levels. So, if you've got depleted serotonin, you'll naturally feel less sleepy. But don't worry, you can hack it – and it all starts in the gut.

GUT HEALTH AND MICROBIOME

If you want to sleep well, you have to think about your gut health, which is something I don't think any GP will talk to you about

* https://www.sciencedaily.com/releases/2019/06/190624173822.htm

if you go to them with insomnia. But it is completely essential. Remember earlier, I explained that 95 per cent of serotonin is produced in the gut? Yep? (Were you listening at the back? Good.) In order to support this, you need to have a healthy gut microbiome.

Your gut microbiome is the community of billions of bacteria and other microorganisms that co-exist in your lower intestines. Scientists are only just now starting to understand how essential our gut microbiota are for health, digestion metabolism, immune function and hormone production.

One way we mess with the variety of our gut microbiome is with eating too much processed food. And by processed, I mean anything in a packet rather than in its natural state. Ready meals, chocolate bars, cereals, packet of crisps, croissants, even those so-called 'healthy' vegan snacks – all of these and many, many more are heavily processed foods that are packed with vegetable oils and other inflammation-spiking ingredients like sugar. Studies have shown that a diet heavy in processed foods actually stops our production of serotonin, because it negatively impacts our gut microbiome and tryptophan levels.

Only a healthy, diverse gut microbiome can produce enough serotonin which then,* as we know, becomes melatonin. So if you're constantly eating sugary, processed foods throughout the day, it will mean you're more likely to suffer with sleep problems (as well as lots of other health issues, which we'll see throughout the book!).

* https://www.wellandgood.com/foods-that-deplete-serotonin/

THE POWER OF FERMENTED FOODS

The good news is there are loads of easy and delicious ways to support a healthy gut microbiome and your serotonin levels. Fermented foods are really well known now for increasing the diversity of your gut microbiome (and the more diverse the bacteria in your gut, the better for you) and has been shown in scientific studies to have incredible effects.*

In one recent study at Stanford University in the States, they compared two groups: one who ate a diet rich in fermented foods and one who ate a high fibre diet. The fermented foods group had much better diversity in their gut microbiomes and decreased inflammation on a molecular level. This was called 'a stunning finding' by one of the researchers and I couldn't agree more.

Fermented foods include things like:

- kefir yogurt
- kombucha (slightly fizzy tea drink)
- sauerkraut
- sourdough bread
- raw cheese
- kimchi (fermented cabbage and other vegetables)

I find the last three things on this list are great for my evening snacky-snacky times. I also love necking down kombucha throughout the day – yes, I may be a little bit addicted, but it's better for you than wine! – which promotes all the good bacteria in my gut microbiome.

* https://med.stanford.edu/news/all-news/2021/07/fermented-food-diet-increases-microbiome-diversity-lowers-inflammation

TOP UP YOUR TRYPTOPHAN LEVELS

Tryptophan is the amino acid that eventually turns into serotonin (which turns into melatonin, I'm sure you've got it by now). We can't manufacture this amino acid ourselves, so we have to make sure our diets include loads of tryptophan-rich foods. Ideally, we would eat these sorts of foods in the evening, a couple of hours before bed.

Foods that are brilliant for this are:

- white rice
- white meat (turkey and chicken – opt for cheaper organic cuts like thighs rather than breast as these are more nutritious)
- grass-fed red meats
- eggs
- nuts
- lentils (even better if you've soaked them overnight before cooking)
- sesame and sunflower seeds
- white fish
- cherries (especially Montmorency ones)
- avocado (not overripe though)
- bananas (ditto)

I boil my rice in bone broth or collagen peptides (10g) too, to boost my levels of amino acids,* which is a really nice little hack. And if I'm feeling snacky and picky at the end of a meal, I'll make myself a tryptophan-boosting smoothie with these ingredients:

* https://www.healthline.com/nutrition/bone-broth#vitamins-and-minerals

- 1 banana
- A glass of raw milk (loads of good bacteria to help my gut microbes)
- 15g grass-fed bone broth powder
- Small tbsp cacao powder (remove if you have high sensitivity to caffeine, as there is a small amount of caffeine in cacao)
- Couple of drops vanilla extract
- Drop or two of stevia sweetener (optional)
- Pinch of sea salt

I whack all of this in the blender and drink it a couple of hours before bed, as it also stops me going down the Mars bar route. You can also take tryptophan as a supplement, which is best taken with carbohydrates. To be honest, I always think food is better than supplements, because you're feeding your gut health too (and food pretty much always ends up cheaper overall). But if you're faced with crappy airline or train food and there's nothing else you can do, you can stack it as best you can with supplements. I'll tell you in a bit about ones that can help with sleep.

EAT HEALTHY FATS TO SUPPORT SEROTONIN

Whatever you do, don't do low-fat food. I've banged on about this before, and I will do until the day I die – no matter what everyone said in the 80s and 90s, low fat is NOT good for you. And when it comes to sleep, low-fat diets can get in the way of that vital serotonin production. We need good fats in our diets as they provide the building blocks of our cells with essential fatty acids, like omega-3. So include as many of these omega-3 rich foods in your diet as you can:

- Mackerel
- Sardines
- Salmon
- Anchovies
- Walnuts
- Chia seeds
- And, if you're feeling fancy, oysters and caviar

Like I've said in my previous book, eating well is about counting chemicals, not calories, and we shouldn't be afraid of fat. I'll go into more detail about how hormones regulate our appetite in the *Why Can't I Stop Eating?* chapter, but just remember, without fat, we would die, simple as that.

SHIFT YOUR EATING WINDOW TO LATER

As well as what you eat, it's about when you eat. When you wake up, you're in a fasted state and that's great, because fasting revs up your dopamine and adrenaline levels. Once you eat, you kick-start the process that elevates your serotonin levels, so you want to push back that release until as late as you can, because serotonin = melatonin = sleepy. This means, ideally, push your eating window to later on in the day.

I know this might sound scary – after all, you'll probably be thinking 'I'll be starving hungry!' But once you shift your habits a little bit, you won't, I promise. Trust me, I know what it's like to feel properly shaky when I was coming off alcohol with the

DTs, and moving my eating window was nowhere near as bad as that! You can utterly change your body clock literally within three days of moving your eating window and not feel hungry at all. So if you normally eat at 7 a.m., try first moving your eating window to 7.45, and then a little later bit by bit. Eventually you would have your first meal of the day around lunchtime, but see how it goes for you.

By eating this way you're not counting calories or worrying about any of that rubbish. You're simply hacking your sleep hormones so they're released when you want them. I don't do well now with eating in the morning – I get very lethargic. So even though it might sound weird and wacky to not eat breakfast immediately, work with it for a few days and see how you feel. We've been brought up to think we need to snack all day long to keep our metabolism going (and yep, I'll go into more detail about this in the next chapter) but actually constantly eating messes with so many things, not least our circadian rhythm. And as we've seen, if we want to sleep well, that's the last thing we want to do.

EAT CARBS AT NIGHT-TIME

As much as you can, it's a really good idea to eat your carbohydrates at night, rather than during the day. They have such a beneficial effect on our sleep hormones – carbs are turned into sugar, which elevates our blood sugar levels and triggers an insulin response. This blunts our cortisol (stress) levels which is ideal for sleep, as well as raising our serotonin levels. Perfect for calming down and chilling out.

You might find, like me, that if you have a sandwich or a

bowl of pasta at lunchtime you end up feeling sleepy in the afternoon. That's your blunted cortisol in action, and it's not always what you want when you have a busy day ahead of you.

So eating carbs during the evening really works well for me. I find when I have them, it puts me in a really relaxed state as it also increases my GABA levels (the neurotransmitter that slows down our whirring brains). So as much as you can, try and eat the majority of your carbohydrates later on in the day, and then give yourself a couple of hours before bed to rest and digest.

Don't get me wrong, I'm far from perfect when it comes to getting this right. Even though I'm a complete sucker for it, I'm trying to break my habit of eating in bed. I started off having my picky bits (sourdough and cheese) downstairs, but me being me, eventually it turned into, 'Oooh, shall we have a cheeseboard in bed while watching a Netflix marathon?' (I mean, *honestly.*) It started creeping in every night, and even though I can sleep OK with a full stomach, like a granddad after Christmas lunch, I know it's not good for me. It's so easy to get into these bad habits, but I'm going to shift it and make sure I eat downstairs in the kitchen a couple of hours before bed. Promise.

SUPPLEMENTS TO SUPPORT GOOD SLEEP

Like I said earlier, it's better and cheaper to have good food sources rather than chucking tons of pills down your neck, for sure. There are so many micronutrients in meat and fish especially that we simply cannot mimic in supplement form – the science just isn't there yet. But there are some supplements out there that can help hack our way to sleeping better. Of course I'm not suggesting you take all of these at once! But hopefully the list below offers a few alternatives. I'd suggest you experiment with what works best for you, trying out one at a time and paying attention to how you feel afterwards.

Magnesium

Magnesium is a mineral, essential for so many of our body's functions, but *especially* for regulating sleep. This is because it activates the part of our nervous system that gets us calm and relaxed, and regulates melatonin production. What's more, it also activates our GABA receptors, which as we know, calm down our whirring brain.* There are many different kinds of magnesium supplements out there, but magnesium glycinate is the most popular for sleep (and is also easier on the tummy than many others). A daily dose can be between 250–300mg.

L-theanine

This is an amino acid that is naturally found in teas and mushrooms and helps our sleep by naturally boosting GABA and serotonin levels. You can get it in capsule or powder format, and I take a 100mg capsule with my coffee every morning. I find it stops the caffeine making me wired and instead I get a lovely calm buzz. I then take 400mg before bed with water.

Taurine

Another amino acid, taurine is found in the body and plays an important part in cardiovascular function, development of muscle and our central nervous system. It's vital for sleep because it activates GABA receptors in our brains and also is

* https://www.healthline.com/nutrition/magnesium-and-sleep#TOC_TITLE_HDR_9

involved in creating melatonin. You can take 500mg–1,500g of taurine an hour before bed. (P.S. I know taurine is in Red Bull too, but I do NOT recommend Red Bull as a sleep aid. Are you bonkers?)

5-HTP

Our bodies turn tryptophan into 5-HTP first before it becomes serotonin and then melatonin. You can boost your levels with supplements of 5-HTP, either in powder or capsule form. You can take 100–200mg an hour before going to bed. (You MUST speak to your doctor about this if you're already taking SSRIs for depression, as it can affect it.)

Vitamin (Hormone!) D

Pretend-vitamin-but-it's-a-hormone D helps you get into a restful state, as it's involved in loads of different sleep functions. Make sure you take a supplement with good fats in the morning, not the evening, as it impacts melatonin production.

Licorice root extract

If you're dealing with a racing heart and panic/anxiety about bed because your adrenals are whacked, licorice root extract can help. It helps support healthy cortisol levels because it contains glycyrrhizic acid (no, I can't say it either), but it means you shouldn't take it at night, so have this in the morning. It comes as a liquid with a dropper which you dilute with a small amount of water.

Zinc

Zinc naturally raises testosterone levels – and low levels of testosterone have been found to contribute to insomnia.* Good levels of zinc have been found to reduce the time it takes you to fall asleep and increase the overall amount of time you spend asleep, too.

Ashwagandha

This is a relaxing adaptogen – herbal remedies that come from plants and help you resist all sorts of stressors. Research has shown that ashwagandha may help people fall asleep faster, spend more time asleep and have better quality of sleep. You can take it as a capsule or in powder form, dissolved in a drink. Warning – can cause adhedonia (this is the feeling of indifference or meh, so use sparingly).

Lingzhi (reishi mushroom)

Another adaptogen, reishi has powerful effects that can support good quality sleep, thought to be down to its compounds that have a sedative and soothing action. Sipping a mushroom tea before bed can be a really nice way to introduce it, or you can take capsules.

3. TEMPERATURE

WHY YOU NEED TO KEEP IT COOL

It probably seems completely bonkers when we've been brought up to think of getting warm and cosy with hot water bottles and

* https://www.everydayhealth.com/hs/low-testosterone-guide/good-sleep-low-testosterone/

electric nana blankets as ideal companions for a good night's rest, but exposure to the cold can be incredibly transformative for improving your sleep. Instead of you getting all hot and bothered and wakeful, the cold will actually help you to produce the right hormones for sleep. I'm sorry, this is the part where I'm going to talk about why you need to get into cold water (I know, I know). But I'm not a sadist, there is some really, really solid science that supports why cold water exposure is amazing.

VA-VA-VOOMING YOUR VAGUS NERVE

The vagus nerve is the longest nerve in the body, running from our brain down to our gut, and it is the main component of our body's parasympathetic nervous system. This can sound a bit confusing, but basically your parasympathetic nervous system is your 'rest and digest' side, regulating all the automatic functions your body does when it is at rest, like digestion. It basically puts our body into its restore mode, slowing our heart rate and breathing down, which is why the optimal function of the vagus nerve is so essential for sleep. (It also regulates things like bladder control, which is why we don't wee when we're asleep!)

By contrast, the sympathetic nervous system controls our 'fight or flight' responses, getting us ready for action during the day.

So, simply put:

parasympathetic nervous system = calming
sympathetic nervous system = arousing

You can activate your parasympathetic nervous system, and calm yourself, with the vagus nerve, which controls about 75–80 per cent of it. And there's nothing the vagus nerve loves more than a bit of cold water exposure.

THE REASON COLD WATER EXPOSURE WORKS

When you expose your body to cold conditions, you increase stimulation of the vagus nerve, which as we've seen, is our body's communication superhighway for rest and digest. Basically, your body deals with the 'shock' of the cold by going 'HUH, what's happening here?' and you get a cascade of lovely parasympathetic hormones to calm you down, which are vital for good sleep.

As I mentioned in my first book, I find cold water works amazingly for me. I often rev myself up by getting really, really hot first sitting in my infra-red sauna pod (about £200 on Amazon) and sweating it all out (mmmm, sexy), and then jumping in a cold shower. I stand there and let the icy water hit the back of my neck so it literally goes onto my vagus nerve at the top of my spine. I find that about twenty minutes after getting out of the shower I get a flood of calming hormones that make me feel lovely and sleepy.

Singing and humming are also known to be great for the vagus nerve, so if you want to ramp this up, you could try belting out a tune in the cold shower.

HOW TO BUILD UP TO IT

Look, I get it. Nobody *wants* to go under a cold shower. Even now, I have to fight the urge that says 'don't do it', because it's human nature to want to stay warm and cosy. But I wouldn't recommend just going straight into a cold shower if you've never done it before, because you're just going to scream and jump out again as soon as possible. Instead, you can work up to it by getting yourself super-hot first.

Basically, we want to trick our bodies into craving the cold. When I first started off doing this, I'd have a really, really hot bath first. You know that point where you can feel the pulse in your head throbbing? Yep, that. I'd put on a podcast and sit there sweating in the bath for as long as I could bear it, and by the end of the podcast it felt horrific and I'd be completely desperate for cold water. So try it – cold water after hot feels pleasurable and you'll actually want to drink that lovely icy water! In fact, if you keep going with it, you'll adapt over time and it won't always feel so cold. There's a bit of magic that happens there.

I do want to say though, this doesn't work for everyone. My friend Lucy tried the cold shower thing and said she just felt completely wired afterwards. We're all completely different, but we've got to try things out to see what works. Give it a go for a week and see how you feel. Try it for thirty seconds under the shower and up to no more than a maximum of two minutes (if you can bear it that long!).

You don't need to do cold water exposure every day – in fact, it's better not to, as you need to keep things random so your body doesn't always expect it (have a look at intermittent dopamine scheduling in the *Why Does It Feel Like I'm Losing My Mind?* chapter). Just a few times a week will really help, and make sure it's at *least* an hour before you go to bed. Or you may find it suits you better in the morning – just experiment and see which works best for you.

TURN DOWN THE TEMPERATURE

Another much easier and less scary hack than cold showers is to make sure your bedroom isn't too hot at night. When you go into deep sleep your body temperature naturally drops so you want to support that. If your room is too hot and stuffy it can negatively impact sleep, making it harder to drop off or even get into that essential REM stage. Think about what it's like when there's a heatwave – it feels horrible and really hard to sleep when you're boiling, doesn't it?

Ideally your bedroom temperature should be around 17–19°, so the first thing you can do is turn down your thermostat or, even better, switch off the heating at night (it's going to save on your energy bills too!). If you can, open the window too, to let some cool air circulate. If you keep your windows closed because of noise or traffic outside, then just invest in a good eye mask and pair of earplugs.

If you find you get really hot in bed, you can cool your core temperature down using cooling mattress pads. These are sometimes called chilly pads and are basically like a really thin lilo that you put under your flat sheet. They're amazing for during the summer months and even for hot flushes (which we'll talk about more in the *WTF Is Up With My Hormonal Cycle?* chapter) even if you're not a bad sleeper, but if you are, I'd recommend having a try.

4. BREATHING

This really is the fastest hack you can do when it comes to improving your sleep – it's definitely a great 'in case of emergency, break this' protocol. Breathing is the quickest way to change your chemistry over and above supplements – you can literally trigger a cascade of hormones without getting out of bed! Breathwork is about getting conscious of our physical state – getting us out of our whirring minds and into our bodies. You don't need to make any big lifestyle changes, or go anywhere in particular, you can literally do breathwork at 3 a.m. lying in bed. Best of all, it's completely FREE.

WHY DEEP BREATHING HELPS US SLEEP

Breathing works because it impacts the parasympathetic nervous system, a bit like cold water does. You can stimulate the vagus nerve with deep breathing; it raises your blood pressure, which sends a message to your vagus nerve, which then lowers your heart rate right down again. By breathing deeply you basically activate those calming, relaxing hormones that you need for sleep.

Someone who's a really amazing advocate for the power of deep breathing (and cold water exposure, too) is Wim Hof. He's a Dutch guy, globally famous as the 'Iceman', and is one of the most incredible motivational speakers around. I follow the Wim Hof breathing method, and if I'm ever in a crisis moment with sleep or ANYTHING, it's the first thing I turn to. I do three to four rounds of Wim Hof, which takes me about 15 minutes, but don't be put off, there's a strange sort of time perception shift when you're in it!

What you do is:

- Take thirty-five very quick, deep inhalations that release very quickly (think of it like the opposite of blowing up a balloon), I do it until I feel slightly dizzy.
- Exhale and hold your breath out for as long as possible.
- When you need to take a breath, take a big, deep one and hold the inhale for ten seconds before exhaling.
- Repeat for three rounds.

After this, I can literally feel a cascade of feel-good hormones. It's the fastest way for me to get serotonin to calm my mood and stop me catastrophising. It has a really similar effect to meditation, which also triggers serotonin and melatonin. I learned about TM (transcendental meditation) in early recovery from alcohol, but I found meditation didn't suit my personality. But if you're dealing with a sleep crisis, I'd recommend trying a few types of meditation as another avenue for you to explore and see if it works for you.

USE GUIDED APPS TO HELP YOU

Of course if you're exhausted and awake in the middle of the night, just the thought of meditation or breathwork can be arduous. I know that feeling, when you just can't be bothered to help yourself – it's bonkers, but it's what we do as humans, so don't start blaming yourself. And feeling like we can't be arsed is even more likely to happen when we're knackered and panicked at 2 a.m.

If that's the case for you, I'd recommend using an app for guided breathwork or meditation. That way, someone is talking you through it and it's *much* easier to do. When I feel 'less-than' I tend to procrastinate, so even though I know how to do Wim Hof by myself, I'll use his app to guide me through it. It's completely normal to feel like you can't motivate yourself, by the way – someone needs to crack the whip for us sometimes. To have constant willpower is just *impossible.* Other apps you could try are Calm, Headspace and Insight Timer – with all of this stuff the key is finding the one that works best for you, and that might take a bit of trial and error.

WHY MOUTH TAPING WORKS

Another super simple hack for improving your sleep is to tape your mouth shut! You want to breathe through your nose NOT your mouth, which so many of us do when we're asleep.

Nasal breathing creates more nitric oxide, which creates increased oxygen and blood flow, lowered blood pressure and improved sleep. When you breathe through your mouth, not only might you snore, but it can lead to high blood pressure and even sleep apnoea (where you actually stop breathing for short periods in the night). Not only that, the dry throat you get from mouth breathing can actually wake you up more in the night, which then triggers that cortisol release, which then can stop you getting back to sleep.

(Matthew sleeps with his mouth open and although he doesn't snore, I can hear a rumble in his throat when he does. The other night I pushed his mouth shut and the rumble literally disappeared [although he did ask me the next morning whether

I'd tried to suffocate him!] I'd like him to mouth tape, but at the moment I'm just using earplugs.)

So, if you know you breathe through your mouth, try mouth taping. You can buy tape that's specifically designed for this purpose, or you can just use no-frills micropore tape instead. It doesn't have to be a full tape over your lips, research has shown that just a small strip of tape across the middle of your mouth, from top lip to bottom lip, will help prompt the body to breathe through your nose.

One good tip is to put Vaseline on your lips *before* you apply the tape, so it's easier to take it off in the morning and you won't rip off any skin (ouch).

A QUICK NOTE ON THE LIMITATIONS OF 'SLEEP HYGIENE'

If you're a poor sleeper, you've probably googled the hell out of sleep advice already, and are likely to have come across 'sleep hygiene' tips. A lot of this is sound advice – stuff like making sure your bed is comfortable, using blackout blinds, not using screens in your bedroom, not drinking caffeine in the evening, winding down before bed with mindfulness, etc.

While there's nothing particularly wrong with any of this advice (and much of it matches up to what I'm saying in this chapter), becoming obsessed with sleep hygiene as a fix isn't going to work. Sleep expert Kathryn Pinkham, of the Insomnia Clinic, has written widely on this topic, and agrees that becoming obsessed with achieving sleep hygiene could actually make things worse. 'Regularly I see people with perfect sleep hygiene sleeping terribly, and actually there's very little evidence that sleep hygiene is a cure for people with insomnia,' she says.*

So, do not worry if you're doing all the 'right' sleep hygiene things and you're still struggling. Here, we're focusing instead on actionable protocols you can apply to YOUR life at all times of the day that directly impact your hormone release.

HOW I HACK MY HORMONES FOR SLEEP – AND HOW TO MAKE IT WORK FOR YOU

I've learned so much about how hormones control our sleep, and I've adapted what I know to my life to make it work for me and my chronological makeup. You literally cannot change how you're wired, and fortunately for our family, I'm a lark, whereas Matthew is a night owl. So, he's great at getting up with the kids in the night if they need something, but try getting him up in the morning – he's a nightmare!

Whereas with me it's different. As soon as I feel that rush of cortisol, I'm up. I put the coffee on, put the news on, I'll go and stand outside to boost my hormones a bit more, get the school uniforms out of the tumble dryer that I forgot to turn on the night before . . . it is total chaos but I find I thrive on it. I do everything I can to stimulate me in the morning, even things like exercise, which I do after getting the kids sorted. If I had a 9–5 job, I'd get up at six and go for a run before the kids got up – I would still stick to the routine that works for me.

* https://www.theinsomniaclinic.co.uk/blog//sleep-hygiene-does-it-really-work

Once I get to midday, I really start to shut down, and by 3 p.m. I'm basically mute! I am *useless* if I exercise in the afternoon – I'm not happy, I'm just wired afterwards, so I've learned it's not for me. In the evening, I do things that promote that serotonin–melatonin release, like having a sauna followed by a cold shower, and eating those snacky, soothing carbs. When I go to bed, I keep the windows open so I have constant fresh air, and put on my totally unsexy eye mask and wax earplugs. I find sleeping with my dogs really comforting, but I know that for lots of people, this would be the opposite of relaxing!

The point is, what works for me is completely individual, and you've got to experiment and explore to find out what works for you. None of us is average and we're all living different lives that put different demands and pressures on us – and of course, our sleep problems will be completely unique! So take what you've read in this chapter and adapt it to your life and your schedule.

GETTING YOUR HOME SLEEP-READY

The first thing to do is make a checklist of your sleeping environment to see what might be sneaking in that might be affecting your sleep. Ask yourself these questions and make the changes if you need to.

What temperature is your house at night? Turn the thermostat down to 18 if it's higher

Is your room completely dark? If not, invest in blackout blinds or an eye mask

Are your windows open? Open them if you can – if this makes your room noisy, wear earplugs

Are there any red lights from TVs? Switch them off at the mains or put a plaster or electrical tape over the light
Have you put your phone in flight mode or switched it off? Radiation from mobile phones has been found to disrupt sleep patterns*
Got an eyemask and earplugs in? Do you look as unsexy as possible? Great!

MY TOP TEN HACKS FOR SCHEDULING HORMONE RELEASE

There's so much information in this chapter and I want you to have a go and try out a few different things, so you can find the blend that works for you. Here's a summary of the main things you can do to help yourself.

In the morning . . .

- **Get some natural light:** stand out in your back garden, on your balcony or stick your head out of the window – whatever you can do to hack your exposure to that cortisol-boosting hormone and get yourself aligned with your natural circadian rhythm
- **Take your sunglasses off:** they'll get in the way of your eye's light receptors and can trick your body into thinking it's night-time
- **Push your eating window to later:** keeping your body

* https://economictimes.indiatimes.com/mobile-phone-radiation-disrupts-sleep-causes-headaches/articleshow/2717605.cms

in a fasted state for as long as possible will help cortisol release in the morning.

At lunchtime . . .

- **Eat fermented food and healthy fats:** a healthy gut microbiome will support your serotonin levels. Bring on kombucha, kimchi, kefir and other foods that don't begin with k . . .

In the evening . . .

- **Wear blue-light-blocking glasses when using screens:** you want to reduce blue-light exposure that will make your body think it's daytime
- **Bananas at night, not in the morning:** keep those tryptophan-rich foods like bananas, oats, lentils and sourdough to later on in the day
- **Dim the lights:** your brain will start producing melatonin in direct response to darkness, so make your home as dim and cosy as possible
- **Have a cold shower or lie on a spike mat:** it triggers a cascade of parasympathetic hormones that support sleep
- **Tape your mouth shut:** breathing through your nose not only improves sleep, but reduces night-time waking
- **Don't watch adrenaline-pumping films at 11 p.m.:** you'll be up all night!
- **Alcohol:** even a small amount WILL reduce sleep quality. Look into Kava Kava as an alternative to alcohol. It's not yet available in the UK, but if you are abroad it

might help with nights out. Or try L-theanine instead . . . Sorry!

YOUR SLEEP-SUPPORTING SCHEDULE, BEFORE AND AFTER

As we've seen, you need to time things accordingly to work with your ancient body clock, not against it. To wrap up this chapter, I want you to take a good look at what you're doing currently in your own daily routine. Be completely honest about what you might be doing at each time of day that could be messing with your sleep hormones. When are you drinking caffeine? When are you exercising (or not at all?) Are you getting outdoors in the morning/at all? Are you wearing sunglasses too early in the day? When and what are you eating? How late are you on social media or your phone? Be brutal, and think about everything and anything that might be adversely affecting you getting optimal sleep.

Once you've thought about your current daily routine, it's your chance to tackle each of those habits that will be working against your natural sleep cycles. Go through what you've read in this chapter and my summary of top hacks, and put into action the ones in that you think will help you. Some of them might be big changes, but I bet you'll find lots of others that are dead easy to do – like getting outside for a couple of minutes in the morning. Don't forget to include a middle-of-the-night emergency protocol. Keep going – things won't change immediately, but you CAN and you WILL sleep better.

DR E – HOW TO TALK TO YOUR DOCTOR ABOUT SLEEP

Sleep is often overlooked in healthcare. I find when we talk to our patients we spend a disproportionate amount of time exploring symptoms a patient might experience while awake, and often sleep isn't even discussed!

Yet sleep is when the body regenerates, and I'm a big believer in supporting our body to heal itself.

There are various schools of thought on the amount of sleep we all need. You may have heard that 8 hours is the ideal, but sleep scientists are coming to consensus on variability in sleep requirements. Our needs vary by genetics, age and what's going on in our biology at the time. For example when we're sick we need to sleep more than usual as all of the energy we create in our body is used up to fight off infections or to heal our tissues.

If you go to see your doctor about any health condition, it is worth discussing how you are sleeping, and what any causes of insomnia may be. Insomnia can lead to a downward cycle of health, inflammation and mental health, so don't ignore it.

Your doctor can help you with poor sleep, both with referral to experts such as NLP (Neuro-Linguistic Programming) practitioners and also with medication where necessary. I would advise you to use prescription medication for sleep issues only with medical supervision, and only at times of crisis – it is highly addictive and can be hard to wean yourself off.

When to go to your GP:

- Racing thoughts: overthinking, anxiety and non-stop thoughts at night (this may be a sign of stress, which has a significant health impact)
- Nocturia: frequent waking up at night to pass urine (this may indicate diabetes, or urinary tract/prostate issues)
- Excessive snoring with moments of breathing pausing (this may be a sign of obstructive sleep apnoea)
- Night sweats (this could be related to an infection/the immune system)
- Cough at night (this could be a sign of acid reflux and even asthma).

When to seek integrative health support:

- Feeling sluggish in the morning and inability to fall asleep at a normal hour; or requiring lots of coffee in the early portion of the day. Adrenal glands dysregulation – test morning cortisol & DHEA, perhaps your cortisol is peaking later in the day (adrenal dysregulation). Could also be an overactive thyroid keeping you up at night. Low progesterone may be a potential culprit in peri/postmenopause (more coming on that later in the book)
- Restless legs at night – although not fully understood, this symptoms has been linked to low dopamine levels
- Twitchy muscle fibres at night – small muscle spasms could be related to mineral deficiencies such as magnesium.

WHY CAN'T I STOP EATING?

IN THIS CHAPTER, WE'LL FOCUS ON THESE HORMONES:

Ghrelin: *the one that tells us we're hungry and need to eat*
Leptin: *the one that tells us we're full*
Insulin: *our blood sugar regulating hormone, triggered by food*
Dopamine: *our reward hormone, which drives our cravings*

Up until a few years ago, I was caught up in a horrific cycle with food. I was always battling with my weight, and I was ALWAYS hungry. In my mind, I was never as thin as I wanted to be, which, of course, back in the 90s was all about that 'heroin chic' super-skinny look, and that was never me.

I was so self-conscious about eating, every time I took a bite of anything I'd feel massive guilt and 'Oh God, I shouldn't really be having this.' I saw my appetite as the enemy, as it seemed to prove to me that I had no willpower when it came to dieting. But as I've come to understand now – and I'll explain in this chapter – I was caught in a trap of eating highly addictive foods that made me overweight, depressed and made it completely frigging *impossible* to control my appetite – aka my dopamine-driven cravings.

This is what a typical daily diet was like for me back in the day when I was fifteen stone and trying to be 'healthy':

7 a.m. Get up, no coffee – it makes me anxious – so I'll go straight on to three Weetabix with oat milk (this new plant-based trend is *much* better for you than regular skimmed milk). Or, I'll microwave a bowl of All-Bran until it's all nice and mushy, and slice a banana over it, as my doctor's told me to have wholegrains and fruit.

11 a.m. Bloody starving. I want a Full English breakfast, but that's really bad for you, right? Instead, I'll reach for a croissant or a slice of toast with Vitalite (less fattening than butter) and jam on it.

1 p.m. I've made it to lunch – hurray! I can finally eat something proper. Why don't I have a salad made with a low saturated fat bottled dressing, some fish and maybe a tiny (OK, not so tiny) bit of brown bread on the side?

4 p.m. Really peckish now. I won't have chocolate, because that's bad for you. I'll put together a healthy brown bread sandwich, with low-fat spread, low-fat mayo, ham and salad.

7 p.m. Finally, it's time for dinner. I've done my 10k steps which is really good (well done, me), so I'll reward myself with a Chinese takeaway to celebrate the end of the day. I'm *exhausted.*

10 p.m. Yet more cereal and toast!

Now I think, no wonder I was exhausted, I was constantly digesting sugar-spiking products! Pretty much everything I ate converted to sugar in my bloodstream:

Weetabix – carbs convert to SUGAR
Oat milk – contains SUGAR and veg oil, plus oats convert to SUGAR
Toast – carbs convert to SUGAR
Jam – pretty obvious what's in that, isn't it?
Salad – to be honest, it was just a gateway to the bread on the side, which is carbs, which converts to . . . you get it, right?

I thought the only way to get energy was to eat carbohydrates, which I had been taught raise your glucose levels and therefore give you energy. (If you're about my age, you'll remember the Ready Brek advert with the glowing orange kid. That message that we need to eat carbs for energy otherwise we literally won't be able to function stayed with me.)

WHY I COULDN'T CONTROL MY APPETITE AND WHAT THAT DID TO ME

In reality, I was caught in the trap of constantly chasing my appetite with food that detrimentally affected my hunger hormones – spiking and crashing my blood sugar levels, releasing the appetite hormones and messing with my satiety signals. I was completely addicted to sugar because I was chasing that dopamine high that I'd lost when I gave up alcohol. I didn't realise it then, but I was biohacking in a really detrimental way, looking for that 'hit' that came from food.

Despite my raging hunger, and eating what I'd been told were 'healthy foods', I couldn't understand why nothing was changing when it came to my weight. After all, I was eating all the 'right' things, according to the doctors. I wasn't drinking, I didn't smoke, I was eating wholegrains and fruit, for God's sake, so surely I should have been losing weight and feeling full? When I went to see my GP about my inability to lose weight, they'd tell me, 'Oh, it's just your age, you've had a few kids now, your metabolism's slowing down. Maybe reduce your calorie intake a bit.'

Right then. I'd try to cut out my snacks, and white-knuckle my appetite until lunchtime. But it was impossible: I couldn't get anything done because the ONLY thing at the forefront of my mind was *when is it time to eat?* I was literally possessed by a demonic chatter in my brain which was CRAVING. It felt so painful, and of course, it didn't bloody work. I'd cave in because I felt as if I was 'starving', even though it wasn't physically possible for me to actually be starving at 5'4" and pushing fourteen stone. But the signals in my brain were telling me to *eat eat eat.*

All this ended up with my mental health on the floor. I didn't lose any weight, so I felt completely 'woe is me, it's not *fair*'. I'd compare and despair with other people, looking at celebrities in *Hello!* magazine and thinking I wasn't as good as them, feeding that beast of self-hatred. I was only thirty-eight, but grossly overweight, stuck in bed with no energy. I was consumed by desperation, that feeling of *surely, this can't be it*? It was that desperation that led me into researching online, which eventually opened up the biohacking pathway.

COMPLETELY CHANGING WHAT I THOUGHT I KNEW

It was when I came across Dave Asprey, famous as the inventor of Bulletproof coffee, that 'fat' started to take on a new meaning for me. Like all of us who lived through the 80s and 90s, I was completely terrified of fat because I'd been brainwashed into thinking it was the devil. But when I started reading proper scientific studies online, all this new knowledge completely blew me away. About how our bodies use fat, about why you cannot rely on calorie-counting for weight loss, and what the hormonal mechanisms are in the brain that cause your appetite peaks and troughs.

Adding MCT oil to my coffee was my gateway to realising everything I thought about food and appetite was wrong. All of a sudden, I was able to get ahead of my cravings by replacing my carbs in the morning with caffeine and good fats. Incredibly, the MCT oil reduced the jitters I also associated with caffeine (eventually, I added L-theanine too, which I've mentioned earlier and I'll go into in more detail a bit later).

With this combination of caffeine and fat, I was able to hack into my brain voluntarily, controlling my appetite and fullness hormones and giving me the creative and physical energy that I'd been lacking for so long. And it didn't take long. Even though my regulated appetite hormones were telling me that once I got up, 'you need a bowl of cereal and toast', after only three days of taking MCT oil, that craving was gone. I broke the habit of a lifetime and it made me realise, wow, there are all these possibilities out there, which led me to find out more and make more positive changes. I realised that I had been fighting cravings for thirty years, ever since I was a little girl.

Now I eat for nutrition, mental health, muscle growth and to repair my body. Now, I eat with gusto. Now, I don't feel hungry between meals. And if you're struggling with that constant, gnawing appetite, that voice in your head that is telling you you're hungry when you've only eaten thirty minutes ago, that feeling of despair if you're trying to lose weight and CAN'T, even though you're doing everything you've been told to do, this chapter will help you. Look, a few years ago, I didn't have a clue. Nobody had told me about how hormones actually work to control our appetite, I just thought it was my lack of willpower that let me down.

WHY THE NORMAL EATING ADVICE IS COMPLETE BOLLOCKS

I've said it before and I'll say it again, *it's not your fault* if you're having trouble controlling your appetite. I see it with my friends and followers all the time – so many of us are caught in an addictive eating cycle, whether that's for things like crisps, chocolate, sweets or 'biscuits, biscuits, biscuits!' (so says Sarah on Instagram).

What really struck me, though, when I asked people online what they find the most addictive and crave-inducing food, is how much it's all geared around SUGAR – and also how much we're all aware that it's got us in its sticky grasp. 'I crave sugar every second of the day,' posted Donna. 'Currently in the midst of a massive sugar addiction . . . bread, chocolate, crisps, wine, potatoes! I know what I need to do, but just can't stop,' added Lindsey. Annette put it bluntly: 'My vice is sugar. Since I stopped drinking about two and a half years ago my sugar cravings went through the roof. I've stopped a couple of times and just end up back worse than I started . . . Sugar is indeed the devil.'

It's so difficult to wean yourself off sugar, when various types of it are stuffed into our food *everywhere.* Overall, I find the hypocrisy of the diet industry completely infuriating. We've been force-fed misinformation for decades about how we should energise ourselves and our lives – being told a low-fat calorie-controlled diet and 'everything in moderation' is the way to go. But these tips are all insanely unhelpful and just *don't work.*

'Everything in moderation' is bollocks because so much of our diet is made up of extremely processed bliss-point foods. These ultraprocessed foods have been proven (and let's be honest, specifically designed) to trigger addictive eating because of how our endocrine (hormone) system reacts to the molecules they contain. To just say 'have everything in moderation' is useless and is probably the most detrimental, patronising approach ever – it won't help you because your body is addicted to these processed foods, and releases hormones that make you want to eat more. You will know that feeling of being unable to have just one biscuit, and then blaming yourself for being greedy or out of control. But's not your fault, it's your hormones.

'Just follow a calorie-controlled diet' is bollocks because we pack our diet with nutrient-poor calories that create completely different hormonal cascades than nutrient-rich foods. A calorie is not a calorie – how our bodies use the calories from good food is completely different to how they use the calories from shit food – and they have different effects on our appetite! Think about it, a small bag of crisps contains *twice* the calories of two eggs. I bet you could eat five bags of crisps in one sitting, but I bet you'd struggle to eat five eggs, not at least ten! What's more, just reducing your calorie intake to lose weight is likely to leave you with a raging appetite.

'Eat low-fat foods' is bollocks because our bodies NEED fat. Eating more good fat actually helps us burn more fat. Low-fat foods are highly processed crap that's produced in a lab, and these food companies replace the fat with chemicals and sugars that have a disastrous effect on our gut microbiome, and mess with the hormones that tell us we're full – so we carry on eating!

This might sound like a rant (and it is!), but what makes me so mad is the lies we are sold. If I went to my GP now with all the information I know, they'd probably tell me *I'm* talking bollocks. It's phenomenal how detrimental some of the most common eating advice is.

Full disclosure here; I was in the middle of doing a personal training course a few years ago, as I thought that was the next step for me, with my love of exercise. The mechanics of the course were super interesting, but then I got to the nutrition part. It was all about controlling your calories, carb loading, eating wholegrains, low saturated fat, eating little and often – literally the complete antithesis of what me and all my followers have found.

After this, I literally had to close the book on it. I thought, 'I cannot possibly, knowing what I know, get behind this. I'm not going to be a PT with this rubbish, so forget it.' I decided that I would just report on the scientific research that's going on, like the work at Stanford University on the gut microbiome, and what works for me and people I know. This outdated nutritional advice doesn't help anybody, and it will just take these big institutions twenty to thirty years to catch up on what I already know, and what I'm sharing with you in this book.

WHY THE FOOD COMPANIES HAVE HACKED INTO OUR BRAINS . . .

We are overfed and undernourished, and we have to realise who benefits from our eating habits – the food companies. Just ten companies own around 90 per cent of the food products that we buy in the shops. It's in their interests financially to have us coming back for more and more, so they hire 'ingredientologists' to make their food addictive, and then market these hyper-palatable foods on us. They *want* us to have a hormonally driven blood sugar low so that we come back to our drug of choice: processed food.

We need to arm ourselves against the major food companies and their constant marketing, plus overt (not to mention subliminal) advertising. Think about it, there's temptation *everywhere*. You can't walk anywhere in a town or city without being assaulted by Costas, Prets, Subways and Greggs, and the non-stop messaging that you should 'treat yourself' with a cupcake.

So, we've been trained to see highly calorific, addictive food as a reward, but then at the same time, have been told that we should control our calorie intake. Of course, one cupcake is never enough, so then we're caught up in a catch-22 from the authorities we trust. What a mess. No wonder women like Claire are caught up in a confusing battle between being 'good' with having 'treats'. 'I'm great all week and then bang – at the weekend I resort to becoming a child, having Friday night treats that extend to Saturday then I spend the next week undoing my behaviour,' she wrote on Instagram.

Who benefits from this messed-up messaging? Not us, not our mental and physical health, and CERTAINLY not our hormonal balance. The only winners are the food companies' profit margins. Cut to CEOs in private jets eating fresh lobster and definitely NOT snacking on what they manufacture in their factories!

. . . AND OUR EMOTIONS!

I bet you can think of loads of foods that give you that comforting, nostalgic feeling because you enjoyed them as a kid, or watched the adverts on telly. I bet you're familiar with that feeling that 'oh, I've had a tough day, I'll go out and treat myself with a cake.' And I bet you're familiar with the feeling of guilt and hunger that comes afterwards. It's something I see so much on my Instagram page.

Melanie posted: 'I seem to just think I deserve a treat after being good for a day . . . my tummy starts to get flatter and my appetite shrinks . . . then I just go into self-indulgence mode till I'm back where I started.' Carolyn also wrote something which really resonated with me: 'I am triggered by feeling emotional and overwhelmed. I then turn to food to numb – I'll sit there and eats crisps, chocolate, cashew nuts, peanuts, Nutella out of a jar, malt loaf with a full slab of butter – anything I can find in the cupboards – and I don't stop until it's gone. Then I feel awful.'

I've been there, and I've got out of it. I want to help any of you dealing with this, so I'm going to explain WHY we can't stop eating, what's actually going on with our hormones that causes this, and HOW you can get ahead of your cravings to sate your appetite and feel great.

WHAT CONTROLS OUR APPETITE

Firstly, let's shift our mindset

Controlling our appetite is not about simply eating less. Firstly, we need to understand that hyper-palatable foods are being pushed on us as 'treats' – but there's nothing in a factory-made brownie that treats us well. We need to take our blinkers off, because in the Western world our health is getting worse. We're all getting sicker, our mental health is on the floor, we've got no energy, we're hitting all sorts of hormonal imbalances and guess what – it's to do with what we put in our mouths on a daily basis.

It's not about getting skinny either – we're all completely different shapes. I don't give a shit if you're 6'2" or 5'1", whether you've got massive boobs or big hips, but to be in a place of danger because you're putting your life at risk due to excess carbs, hidden sugars and inflammatory oils because of the profit margin at these huge companies is *not fair.* And then being told it's our fault because we've 'got no willpower' to stop eating, when we're being served something that's as addictive as cocaine in the form of a plant-based granola bar, is bonkers!

What we're going to do in this chapter is get ourselves into homeostasis – a state of balance between our physical, chemical and biological functions. We're going to override the hormone-driven craving mechanism which underpins not only our appetite but so many chronic illnesses that are prevalent today. We're going to feed the brain, and the body will follow. Once you stop the addictive eating of highly processed food, I promise you'll see an amazing shift in your health. It's nothing to do with willpower, it's to do with hacking your hormones to support yourself. Let's find our own status quo so we can live long and happy.

How hormones influence our appetite

Hormones are in complete control of our hunger levels – and when they're released and at what levels dictates what food (and how much) we reach for. Like with everything hormone related, there are so many at play when it comes to appetite, but these are the major ones we're going to focus on:

Ghrelin

This is the hunger hormone. Ghrelin is mainly produced in the stomach, and its primary role is to regulate our appetite by sending a signal to our brain that it's time to eat. Think about what happens when you walk past a fast-food takeaway and smell chips – you often get a feeling that you're hungry, right? That's ghrelin being released.

Ghrelin levels vary throughout the day, and can be triggered not only by food smells, but also our cell's body clock. If you eat at very regular times of the day, your body will learn to produce ghrelin at these times, because it's 'expecting' to eat. What's more, lowered blood sugar levels will also trigger ghrelin, which is why we feel hungrier after eating sugar-spiking foods. It's the *drop* in blood sugar from the high level that tricks ghrelin into being released, not the fact that we're actually in real need of food.

Leptin

Leptin is the satisfaction hormone which basically tells us 'right, you've eaten enough now, you're full.' Leptin, which is produced by our fat cells, sends signals to the hypothalamus (in the brain) which produces a feeling of satiety. Maintaining our leptin levels is really crucial for regulating our body weight as it controls how we burn our calories, too.

Leptin evolved to stop us humans from starving or overeating, but modern processed foods really mess with our leptin receptors in the gut, because they strip the mucosal lining that contains these receptors, and so the leptin can't do its job properly. What's more, leptin resistance – where even though your body is producing a lot of leptin, the brain doesn't see it – is now believed to be one of the main sources behind obesity. If your brain isn't receiving the leptin signal, it thinks you're starving, and so you eat more.*

Insulin

Insulin is the hormone that regulates our blood sugar levels, which, as you'll see, is massively important when it comes to

* https://www.healthline.com/nutrition/leptin-101#leptin-resistance

our appetite. It's created in our pancreas and is released when our body breaks down food into glucose (sugar), which our cells use as energy. As the glucose moves around our cells, it triggers an insulin response which then brings our blood sugar levels down to normal levels.

The problem with our high-carb, highly processed diets, is that they all convert to sugar in our bloodstream, which produces insulin spikes all day long to try and bring our blood sugar levels down to a safe level again. This drop in blood sugar then triggers ghrelin, the hunger hormone that tells us 'I need some more food!' Because of this, we end up eating more of these high-carb, highly processed foods, and create an insulin spike all over again, so we're now in a vicious cycle of peaks and troughs. Our bodies don't need all this energy, and so the excess glucose that isn't used by our cells will be converted and stored as fat.

For me, keeping my blood sugar levels balanced means I avoid getting those lows which make me feel starving hungry. What's more, if we're constantly snacking we're repeatedly spiking our insulin levels, which can lead to pre-diabetes, which can lead to type 2 diabetes, which can lead to some horrendously serious illnesses. So in short, we want to keep our blood sugar as balanced as possible.

A NOTE ON DIABETES

Type 1 diabetes is a genetic condition where the body doesn't produce enough insulin, causing high glucose levels in the blood. This can result in horrific health problems, and so people with type 1 diabetes need to inject insulin every day in order to survive.

Type 2 diabetes is where your blood sugar levels are too high because the insulin response isn't working properly. It's the most common sort and is brought on by our high-sugar, refined carbohydrate-heavy diets. The number of people with type 2 diabetes is rising rapidly, and it's estimated that more than 5 million people will have type 2 diabetes in the UK by 2025, which is 7.5 per cent of the population.* Scary – but the good news is that it is reversible through diet and lifestyle changes.

Dopamine

Ahhh, dopamine, it gives and it takes. This is our desire hormone but is also the source of addictive behaviour. It's a neurotransmitter, produced in the gut and in the brain and is a great motivating hormone that controls all sorts of behaviours. It used to be thought of as just the 'pleasure' hormone, but it also motivates you and activates multiple cells and actions. Dopamine will make you turn on the TV and put on the kettle, knowing that the reward is coming in the form of a show and a cup of tea. It controls our movement and speech, and it regulates our reward pathway.

Dopamine's important when it comes to appetite because it controls our cravings. It's the motivating hormone that back in prehistoric days, would have made sure we got out and foraged for our food. But now we have dopamine triggers everywhere; from getting 'likes' on our phone, to hunting out that croissant (more on that on page 76) that give us a dopamine high.

I'm naturally low on dopamine which is why I seek it out so much. I used to chase the dopamine high with alcohol,

* https://www.medicalert.org.uk/news/2020/12/03/diabetes-and-lifestyle

and then I replaced it with sugar and addictive eating. Those dopamine cravings are horrendous, which is why you don't want to white-knuckle it. I bolster my natural dopamine levels now with running, upbeat music, light exposure and all sorts of supplements – as well as scheduling my eating window and what I eat, which we'll explore in a moment.

How they all work together – aka The Croissant Pathway

All these appetite hormones work together to influence how, when and what we eat. The best way to illustrate this is to describe what's going on when you eat a croissant – and it starts even before you've got your hands on it! Imagine an average morning goes like this:

- You wake up too late for your regular breakfast, and so think about grabbing a croissant before work this morning. Dopamine is released at the thought of this 'reward', motivating you to get out of bed and get ready so you can get to the croissant. You usually eat at 7.30 a.m., so ghrelin is released in anticipation of your regular breakfast and you start to feel hungry.
- On the train to work, you can't stop thinking about the croissant because ghrelin is gnawing in your stomach. It's raining when you get off the train and you don't have an umbrella, but the dopamine drive (get that croissant!) is strong enough to get you to the café anyway.
- You buy and eat the croissant and enjoy the pleasurable sensation from dopamine as you've achieved your 'reward'.

- The carbs convert to sugar in your blood, triggering an insulin response and also inhibiting your leptin receptors – so, even though you've just eaten, you don't feel full. You debate getting something else, but you've got to get into the office first.
- At your desk, insulin does its trick and brings down your blood glucose levels. This drop triggers ghrelin again, so your appetite comes back. You try and fight it, but it's no good – at 10 a.m. you're in the office kitchen raiding the biscuit tin.

Sound familiar? Or something like this does, I bet. Our hormones are driving all this behaviour, but the problem is, that dopamine hit soon becomes routine and so we end up eating more and more to get that same feeling. Ultimately, we start to biohack really detrimentally, chipping away at our baseline dopamine levels and becoming controlled by our cravings.

Before you know it, you need two croissants to get the same hit of dopamine and it becomes routine. So what started off as a dopamine-driven occasional treat becomes a beige everyday habit, and one with serious consequences for your long-term health. We need to feed our brains as well as our bodies, because enthusiasm for life comes from the brain, not from a croissant.

What causes addictive eating

What's happened with obesity levels over the past thirty years has compounded everything we all know, deep down: highly processed foods cause addictive eating. (In the US in ten years' time 50 per cent of the population are going to be morbidly obese and you're telling me it's *not* down to processed food, that didn't exist pre-1950s?) Our Western diets are stacked with foods that come in a packet with lists of unrecognisable ingredients. They're created in labs by scientists, not chefs, and are designed to hit our 'bliss point', which is a blend of refined sugar, refined salt and fat from vegetable and seed oils that we find irresistible. Literally.

When we eat processed foods, we completely mess with the hormonal pathway that controls appetite, as we've seen in the croissant example. We continually spike our blood sugar, causing a hormonal cascade that leaves us hungrier than ever, and reaching for the next processed 'treat'. They make us fat and miserable and inflamed and very, very ill, but they're made to be hyper-palatable, i.e. super tasty, so it's a horrendous trap.

I'm completely anti processed foods, because I've experienced first-hand how shit it feels when you're addicted to them. The lack of actual nutrients in some of these foods is abysmal, and we are hard-wired to seek out certain nutritional targets in what we eat (a combination of amino and fatty acids) This means if we don't hit these nutritional targets in what we eat, we're simply going to reach for more food!

That's why you're always going to go for a bag of chips after eating one of those fake-meat Impossible Burgers, because they're full of chemicals and you're not hitting the nutritional targets your body needs, like B12, heme iron and all the other vitamins and minerals you get in natural food sources like meat, fish, dairy and eggs.

So there's no blame here at all – if you can't stop eating processed foods, it's not your fault, you're not 'weak' and you don't 'lack willpower'. These foods are designed to make your body crave more on a chemical level. As a society, we have been force-fed highly addictive substances and utterly rubbish advice.

The ingredients to watch out for

The first good thing you can do to start regulating your appetite hormones is to READ THE INGREDIENTS on any packet of food. Don't take the packaging design as an indication of how healthy or unhealthy something is, literally read the ingredients list and if any of the below are in there, don't eat it. Remember, by piling our diets full of these ingredients, companies are hacking our body chemistry for their profit. And also remember – if it comes in a packet, it's not as good for you as the food that doesn't, like fish, meat and fresh vegetables.

Vegetable and seed oils

If you read my last book *It's Not A Diet,* you'll already know why vegetable oils are my number one food enemy. They're just a disaster – an inflammatory shitstorm for your body and brain. This is the first thing you need to get rid of, over and above even sugar.

Not only do refined vegetable oils like sunflower and rapeseed or that generic 'vegetable oil' cause inflammation, because they're full of omega-6 fatty acids (which are the ones you *don't* want a lot of), they mess with the leptin receptors in our gut mucus. When we eat vegetable oil in any form, it strips the essential gut mucus, so the leptin receptors retract and we never get the signal to our brain that we've had enough. So, if you eat a lot of things containing vegetable oil (and believe me, *so many* processed foods do) not only are you running the risk of developing inflammatory diseases like diabetes, arthritis or cancer, you'll go back for more after half an hour.

It's worth remembering that vegetable oil was originally used as a detergent for cleaning farm machinery (ewww) – and it was only later on in the twentieth century that it began being used as a cheap ingredient in processed foods. Originally it was sold to us as a 'healthy' alternative to butter, because it's unsaturated, as opposed to saturated fat. Over the last few decades, consumption of vegetable and seed oils has *rocketed* as our diets have become more processed.*

All these cheap veg oils – which are in everything from crisps to crackers, ready meals to biscuits and so much more – contain high levels of something called linoleic acid. It now makes up about 8 per cent of our daily calorie intake and researchers believe this is more than we've been evolved as a species to handle. Eating too much linoleic acid has been shown to mask the sense of fullness we get from leptin, and also increase the size of our fat cells. What's more, it also stops other 'good' fatty acids (like omega-3) being absorbed by the body. So while we're getting fatter and hungrier, we're also getting unhealthier.

If you're still in doubt, there's a well-known cleaning product called The Pink Stuff Paste that proudly boasts it's 'plant-based'. What that means is it's a vegetable-oil based cleaner! Think about it; if you can get rid of rust on your garden furniture with veg oil, then what on earth is it doing to your sensitive gut lining? I interviewed Dr Cate Shanahan (who blew the lid on the seed oil scandal) who said she breezed through menopause AND grew an inch in height, as soon as she removed these toxic ingredients from her diet.

Emulsifiers

Emulsifiers are chemical additives in our food which bind ingredients together like oil and water. They're added to

* https://www.forbes.com/sites/realspin/2015/09/29/could-so-called-healthy-vegetable-and-seed-oils-be-making-us-fat-and-sick

everything from salad dressings, bread, sauces, margarine and ice cream, because they keep things creamy-looking. It's an aesthetic thing rather than something we actually need – they don't even have a flavour! Emulsifiers are actually a bit like a glue. Look out for names like soy lecithin, carrageenan, mono- and diglycerides in the ingredients list, as it can sometimes be tricky to spot emulsifiers.

Emulsifiers in highly processed foods also inhibit the gut's ability to detect what's actually in your food, and therefore stop them releasing leptin, the satiety signal. They do this by shutting down a hormone called CCK (cholecystokinin), which is in your gastrointestinal tract. Basically, CCK triggers leptin signalling, but emulsifiers shut this down by causing damage to the mucosal layer so they don't detect what's in the food. That means the leptin doesn't rise, and the signal that you're full takes longer to reach the brain.

Sugar

You might think your granola bar snack is healthy because it says 'no added sugar' but sugar takes on many different, sneaky guises. All these are just different types of sugar:

- fructose
- glucose
- *any* syrup, e.g. rice syrup
- barley malt
- fruit juice concentrate
- sucrose
- agave nectar
- dates
- honey

Hidden sugars in foods create more appetite by messing with our insulin response, so you eat more. Fact. And it's not a coincidence that so many of my followers talk about being addicted to sugary foods, especially sweets. 'I'll eat a whole bag of gummy bears without drawing a breath. Then feel sick, but do it again the next day,' said Louise. Dawn added; 'I crave jelly snakes every night about 9 p.m. It got to the stage that I was buying them while waiting for the school bus and eating the whole packet in the car on the way home!'

But it's not just the obviously sweet foods we need to watch out for – hidden sugars are added to pretty much all processed foods, even savoury ones like soup (honestly, take a look at Heinz Cream of Tomato soup! It's a sugar bomb).

A note on depression and addictive eating

It really bears mentioning here too that you may be overeating because you're depressed. I will go into more detail later about how hormones affect our mood in the chapter *Why Do I Feel So Low?*, but there is certainly a link with addictive eating.

When I was depressed, overeating was a vicious cycle. My pleasure–pain dopamine balance was knocked out of whack, so I reached for things to boost my emotions and 'feel' something (although I didn't understand this at the time). My weak spots were pizza followed by ice cream, and then I'd cycle back through the sweet–savoury cycle. Some people reach for drugs, or alcohol – which by then I'd given up – so processed foods became my drug of choice. When you're depressed, you're likely to have higher cortisol levels and low serotonin, so you'll be seeking out carbohydrates, which are always in these highly processed foods.

If you're dealing with depression you probably have little or no emotion, so you will go for the foods that made you feel safe or happy in the past. And this is understandable, but you need to look at the ingredients and ask yourself is it triggering an addictive response? After a bowl of cereal are you likely to say after twenty minutes 'I'm satisfied' or will you be back hunting for more? I know it's very difficult to dig yourself out of this cycle but remember, you're biohacking detrimentally. By avoiding these foods, you'll be starting to biohack in a positive way, and your appetite will stop taking over your life.

BUSTING FIVE FOOD MYTHS THAT WILL HAVE MADE YOU OVEREAT

1. Plant-based does NOT mean healthy

Right (cracks knuckles), this is the bit where I'm going to lay into this vegan trend. I have a massive issue with veganism in general, because I believe most vegans will be in nutrient deficiency. There are so many vegans that go back to eating meat, because they're prediabetic, have no energy, have piled on visceral fat around their liver because their diets are too carb-stacked, the list goes on. In my opinion, being vegan is not a sustainable lifestyle for most people, and we've not evolved as a species to eat like that.

The reason I think veganism has become so popular recently is because so much of vegan food is actually processed junk food. People say, 'Oh, I'm plant-based' as if it's a badge of honour, but in actual fact, you're just totally fast-foods based. All of those vegan burgers, sausages, fake meats, fake cheeses and vegan spreads are just stuffed with inflammatory chemicals and lack any real nutrients.

I am sure there are vegans out there who are eating a balanced diet of unprocessed foods, but I have found there are so many vegans who survive on chips and veggie sausages and wonder why they're tired, hungry and overweight. It's because you're not giving your body the nutrients it's evolved to survive from! Your gut is completely inflamed and you're not able to produce the hormones your body needs like serotonin and dopamine. How exactly do you expect to beat Mother Nature?

I am not here to tell anyone that they should or shouldn't be vegan, be paleo or be carnivore, I'm just saying be kind to yourself. Processed plant-based food is still just junk food. If you don't want to eat an animal for ethical or religious reasons, fine, have an egg. It is the source of life. It's got every single ingredient there to grow life. And at least you're ticking all the nutrient boxes.

Whenever I see 'plant-based' I read 'chemical shit-storm'. This is green washing at its finest. If anyone thinks that a vegan or vegetarian diet is death-free then think again. Just watch *Watership Down* if you think that animals aren't slaughtered to make way for huge tracts of crops. Teens are particularly vulnerable to the plant-based rhetoric, but low levels of heme iron and B12 will be detrimental to their mental health, no question about it.

2. A McDonald's burger is better for you than a vegan burger

I'm not advising you to run down to MaccyD's and order a Big Mac – McDonald's is not exactly the healthy option, and most of the things they sell are completely addictive, sugar-and-insulin-spiking, processed foods. When we say, 'Ohhh I really fancy a McDonald's,' what we're really saying is, 'I fancy the bun, I fancy the mayonnaise mix, I fancy the ketchup,' *not* the burger.

The burger itself is not a problem, it's everything around it. Because actually, just the burger is infinitely better than a so-called 'healthy' vegan burger. If you can restrict yourself to just eating the burger, then knock yourself out! Just look at this ingredients list:

McDonald's burger	*Impossible vegan burger*
100% ground beef	Water
	Soy protein concentrate (inflammatory)
	Coconut oil
	Sunflower oil (inflammatory)
	Natural flavours (whatever these are)
	Potato protein (another emulsifier)
	Methylcellulose
	Yeast extract
	Cultured dextrose (another word for sugar)
	Food starch modified
	Soy leghemoglobin (inflammatory)
	Salt
	Mixed tocopherols (antioxidant)

Soy protein isolate
Vitamins and minerals (zinc gluconate, thiamine hydrochloride [vitamin B1], niacin, pyridoxine hydrochloride [vitamin B6], riboflavin [vitamin B2], vitamin B12 [synthetic B12 aka cyanocobalamin – from cyanide!]).

So, there you have it. A cocktail of inflammation and insulin-spiking ingredients and lab-created vitamins that mess with your gut microbiome, hormone production and leave you feeling hungry. Yum! Now of course the source of the meat also matters and, depending on where in the world your beef was raised, non-organic can be pumped full of hormones and grain, so it's not necessarily the best meat but it's sure as hell better than the ultraprocessed alternative. In the UK and Europe we currently have the most stringent laws on the planet for raising cattle, so bear that in mind when you buy.

3. Snacking makes you hungrier

As a society, we have this mania around little-and-often snacking, which has been forced down our necks by – guess who – the food companies, because snacks aren't free. Just think back a few generations to our grandparents, they didn't snack between meals and they didn't die! Now there's this mentality around snacking, 'Oh my god, I've not eaten for a couple of hours!' You know what, you're *fine,* Pauline!

The fact is, we can't survive for more than a few minutes without air, and not more than three days without water, but in some very, very extreme cases, people have survived without food for months at a time. There was a famous case back in the 1960s where a very obese Scottish man, Angus Barbieri, lost 125kg by not eating for 382 days (although he did have vitamins, yeast and various drinks)!* I'm not suggesting not eating for over a year OF COURSE, I'm not a maniac, but we need to reprogramme this obsession we have with snacking being 'good' for our metabolism. In fact, if we snack regularly, we're constantly spiking our blood sugar and insulin response, and end up burning carbs for energy, not our fat stores. Then that drop back down in blood sugar triggers our hunger hormone again.

Eating little and often does nothing for our waistline. It makes us store fat and become brain foggy and in a constant hunt for food because we're dropping our blood sugar constantly. It's so much better to eat more at fewer intervals rather than spreading out our food intake throughout the day with snacks every couple of hours. We could be eating exactly the same amount of calories, but we'll feel *less* hungry if we keep our blood sugar steady by giving ourselves a few hours without eating.

(Side note: it's also been proved to be really medically beneficial to intermittently fast. A Japanese scientist called Yoshinori Ohsumi won the Nobel prize in 2016 for his work on autophagy, which is what happens to your cells when you fast. Autophagy can help get rid of toxic proteins in the cells [which can turn into neurodegenerative diseases like Alzheimer's], recycle cells, prompt cell regeneration and reduce inflammation.)†

* https://www.diabetes.co.uk/blog/2018/02/story-angus-barbieri-went-382-days-without-eating/

† https://www.nobelprize.org/prizes/medicine/2016/press-release/

4. Fat is good for you

I want to get this message into your head: fat is a fuel! If you consume carbs, you burn carbs. But if you consume fat, you burn fat. I have *so* much fat in my diet now, because of what I've learned. Low-fat diets are horrendously bad for us and just make us fatter, because our body is burning carbs constantly and storing the excess energy as fat. In fact, fatty acids like omega-3, are super-important because our body is programmed to achieve certain nutrient targets in our food.

When you're genuinely hungry and producing ghrelin, the body is craving a certain amount of amino acids and fatty acids, and it will seek them out no matter what you're eating. So, for example, if you eat a slice of toast, it will have a certain amount of amino acids in it, but it won't hit all your fatty acid targets because it's not a complete food. So, your hormonal cravings will just cause you to carry on eating until it hits this amino and fatty acid target. However, if your food is nutrient dense, which good fats are essential for, you'll become satisfied quicker, and your leptin response will tell you to stop eating.

Instead of worrying about low-fat foods, and calorie intake, you need to start reframing your brain to understand that clean fat is really good for your health and appetite. You can do this either by eating good fats like avocados, eggs, oily fish, animal fats and grass-fed butter – or by cooking or adding good fats to your food, such as:

- MCT oil
- olive oil/extra virgin olive oil (by adding it to your salad you make all the goodness in the salad more bioavailable, as fat facilitates nutrient absorption)

- flaxseed oil (I have a tiny bit of that in a smoothie sometimes)
- avocado oil
- fish oils
- pumpkin seeds

5. Your body needs cholesterol

Cholesterol is a waxy substance found in every cell in our body – and it's vital for producing certain hormones including oestrogen, progesterone and vitamin (HORMONE!) D, along with all your sex hormones! It also makes the liver acids that absorb excess fat during digestion.

If you grew up in the 80s and 90s, you'll remember the hysteria around cholesterol in the media, and even now cholesterol is still being described as something awful, and we're told that we should reduce it at all costs. But it's not as basic as 'high cholesterol is bad for you and you need to go on statins'. Doctors are over-simplifying cholesterol and making us scared of it. The truth of it is that we *need* cholesterol and it's paramount to healthy hormone levels.

We have two types of cholesterol: HDL (high-density lipoprotein) and LDL (low-density lipoprotein).

HDL is referred to as 'good' cholesterol as it picks up excess cholesterol in your blood and takes it to your liver, which flushes it away, and therefore protects your heart from disease. LDL cholesterol is the 'baddie', which can build up in the walls of your blood vessels, potentially forming a clot, and subsequent heart attack or stroke. So, if you know your cholesterol level, you need to differentiate between the HDL and LDL levels – they are NOT the same!

You've got to reframe your perception of cholesterol when it comes to eating, because eggs – which are high in cholesterol – are one of the best things to eat to regulate your appetite and blood sugar hormones, as they're so nutrient-dense. I have eggs every day – it's sugar and carbohydrates that increase bad cholesterol, so do not blame the humble egg. We've been eating them as humans for thousands of years and then suddenly the *Daily Mail* says they're bad for you? It's absolutely preposterous. The American Heart Association now report that cholesterol is 'not a nutrient of concern for over-consumption'*, yet the word has still not spread through the media, let alone to most doctors.

HOW TO REGULATE YOUR APPETITE AND DITCH ADDICTIVE EATING

Feed your gut, feed your brain

I talked in the sleep chapter about how important the gut microbiome is, and it's just as vital here in the food chapter, as you've probably already worked out. We can heal our guts within just a couple of weeks by eating the right things, populating the gut with good bacteria and ditching processed foods. Our billions of gut microbiota are vitally important for healthy hormone production, which sends the right messages to our brains at the right time.

We need to keep our brains healthy, because that's where we decide to reach for either our trainers, or the bag of Square

* https://www.ncbi.nlm.nih.gov/pmc/articles/PMC6024687/

crisps! It all goes back to the brain–gut axis, and ensuring our hormones are doing the job right, and managing our appetite in a healthy way.

What I do every morning to manage my appetite hormones

I am an addict by nature so I will ALWAYS press the fuck-it button unless I get ahead of my cravings. I have to think ahead and not leave things to chance. It's like when I was in early recovery from alcohol, would I have gone and sat in a pub? Of course I wouldn't. If you sit in a hairdresser's long enough you'll get a haircut and if I stay in the kitchen without planning out what I'm going to eat or drink, it'll all go to pot. We know ourselves, we know what triggers us, even if it's as simple as looking at a clock and thinking 'Oh, I should eat now'. So, I'll always try to get ahead of the curve.

Unlike a lot of biohackers, who say they wait an hour before drinking their morning coffee, no, that's not me because I've got three kids to get out of the door and I need the caffeine immediately! As soon as I get up, I'll have water (if I've been organised, I'll have a jug already in the fridge, but most of the time the reality is it'll be tap), squeeze some lemon and put a pinch of salt or some decent electrolytes in it and I'll neck it while the kettle is boiling. Electrolytes give the brain the magnesium, potassium and sodium minerals it needs to work properly – i.e. for the electricity to flow easily.

While I'm making the kids breakfast, I'll take my L-theanine and get my coffee on the go, chuck in my keto powder (with MCT oils) and grass-fed butter which I found really reduces my morning bloat. I have a sweet tooth, so I add a bit of stevia in here too, but stevia doesn't spike my insulin like sugar would. Some recent studies are suggesting that it is actually anti-inflammatory! Winner. (There is some debate about whether stevia messes with your gut microbiome, so the jury's still out on that one. But for me personally, if it's keeping me off the white stuff, then it's a net positive and I'm happy to keep using it.)

The MCT oil powder in my coffee really parks my cravings, keeps me in fat-burning mode (which I've entered during sleep), and stops ghrelin in its tracks. Otherwise I know I'd be picking at the kids' bacon! And I know people say, 'Oh, I don't have time to cook real food first thing,' but you *do.* I always boil four eggs in the morning and it takes no longer than making my kids cereal and toast. Asa and Jude have a chucky egg in a cup, Luxx has bacon, and the extra ones are for me for later.

Like I explained in the *Why Can't I Sleep?* chapter, I then do all my cortisol/adrenaline/dopamine-boosting activities in the morning and early afternoon. When I get hungry, I'll then have those boiled eggs with some salt – either on their own or before anything else. By eating those first, I can guarantee I won't then reach for a bag of crisps when I go to the petrol station at 4 p.m.! Plus, I've extended my eating window, kept my blood sugar balanced and managed my hunger hormones so I can enjoy eating real food later on in the day.

FOUR WAYS TO REGULATE YOUR APPETITE WITHOUT PAIN

Not eating for a few hours can be HELL when you're dealing with the twin beasts of ghrelin and dopamine – I understand, I have been there and then some! But you can do it by reprogramming your neurobiochemistry, to get you ahead of the cravings before they pull you towards Pret. You can avoid the blood sugar peaks and troughs and get free from the horrendous appetite chatter in your brain.

I know it can be a hard sell, me telling you not to eat your Weetabix and instead put some powder or oil in a coffee. But think of it like this (which really helped me) – it's not like you're *never* going to eat it again, it's just for right now, so you can start to bring your hormones back into balance.

It's amazingly liberating to conquer your cravings. You'll have changed your neural pathways, so you won't go back to your old habits once you've done it (and even if you have a day when you fall off the wagon, *that's fine,* I certainly do!). It's completely possible and I promise you will not feel hungry, but more energised, creative, sated and joyful.

1. Move your eating window to manage insulin and ghrelin

If you read the last chapter on sleep, you'll have seen what I said about your cells being on a body clock, releasing certain hormones at certain times. And that's completely true when it comes to your appetite as well.

Ghrelin releases in accordance with your regular eating times, so you need to start shifting when you eat to prevent that dose of ghrelin that will tell you 'I'm hungry!' It's also sparked by a drop in blood sugar, as you'll have read earlier in this chapter. As we know, that drop is caused by eating glucose-spiking foods, which trigger an insulin response and therefore cause the drop. So it's vital to manage your insulin response to manage your appetite, and you can do that by shifting your eating window to later in the day, bit by bit.

Your body literally won't notice it if you move it by forty-five minutes each day. So instead of eating at 8 a.m. if that's your regular time, eat at 8.45 a.m. After a couple of days, shift that to 9.30 a.m. and so on. You can do that easily over time and you'll stop feeling so hungry when you get up in the morning. Promise!

I'm somebody who used to get up at 7 and have three Weetabix immediately because I was 'starving' (of course I wasn't, but you know what I mean). I'm not asking you to fast for a year like that Scottish bloke, I'm asking you to gently push back your eating window so you schedule your hormones right, and start to take control of your appetite.

2. Keep control of cravings *before* you eat (plus the incredible new natural supplement that reduces appetite)

There's nothing worse than trying to battle your dopamine cravings, but you can get ahead of them with supplements and ingredients to add to your drinks or to have before you eat.

L-glutamine

This is an amino acid that I used when I was coming off sugar. You put a spoonful of this powder under your tongue for thirty seconds, it gets into your bloodstream and it dampens down the sugar craving (but tastes of talc!)

MCT oils

Add some MCT oil or powders to your morning coffee or green tea, and that will hit your brain with noradrenaline and take you away from cravings. MCT (medium-chain triglycerides) contains amazing fatty acids derived from coconut oil. They're quickly absorbed into the bloodstream and sate your appetite because they contain good fats and get your brain and body switched on.

Collagen

Collagen contains so many amino acids, which as I've mentioned, is what our bodies are craving when we're hungry. So if you want to manage your cravings, add some collagen powder to your morning fatty coffee. It's a good thing to have before you go out for dinner, too, and will help you have more power against the bread roll!

Apple cider vinegar

This is such a great and simple hack, and what's more, it's dirt cheap. Have a spoonful of apple cider vinegar before you eat and it will kick off your digestive enzymes, slowing the breakdown of carbs in your gut. It also dampens down your insulin response, managing your blood glucose levels better (i.e appetite!). You can take this in a glass of water if you don't like the taste of it neat.

Berberine

This is one of my absolute favourite new discoveries, and it's not an exaggeration to say that it's taking the medical world by storm too. Berberine is a completely natural plant product (a bioactive compound derived from tree bark, if you want to get all fancy) and has been shown in studies to reduce the secretion of leptin, our appetite hormone. It's a completely safe, natural supplement that reduces our blood sugar levels and brings the insulin response right down.

Berberine works by activating an enzyme in our cells called

the AMPK enzyme, our metabolic 'master switch', which is found in organs all over the body. It works in so many different ways: decreasing insulin resistance, helping our body break down sugars, decreasing the sugar production in the liver, slowing the breakdown of carbs in the gut and even increasing the number of good microbes in our gut. It's also shown to increase dopamine production from the gut too (via microbiome regulation).

What this means for YOU, is that *berberine will help you stop eating too much.* If your blood sugar levels are stable, your hunger pangs are reduced. If you're going to have your Chinese takeaway, or a high-carb meal, then you'd do well to have some berberine first – it'll stop you going back to the fridge for picky bits later on! I could really have done with some berberine back in the day when I was overweight, but even now, if some major catastrophe happened and I just thought fuck it, and wanted to self-sabotage with doughnuts (as I do every now and again) it would be very handy to have berberine first.

What's really exciting on a global scale with berberine, is that it's been shown in studies to be as effective as pharmaceutical drugs in helping obese people lose weight and reverse their type 2 diabetes. At the moment, people will be prescribed something called metformin to treat their type 2 diabetes, which has all sorts of side effects like heartburn, bloating, headaches, metallic taste in the mouth, extreme tiredness – the list just goes on. In one study, people who took berberine lost 3.6 per cent of their body fat, going from obese to overweight in just three months. Other studies have shown a conclusive anti-diabetic and anti-obesity effect* – it really is a wonder supplement. Look out for di-hydro berberine as it is more bioavailable.

* https://pubmed.ncbi.nlm.nih.gov/33415147/

Korean ginseng

Improves glucose and insulin regulation thanks to it containing ginsenocides, which are also inflammatory.

Holy basil

Is an adaptogenic herb which helps to combat stress and promote blood sugar stability

3. Change the order you eat your food

This is such a simple and incredibly effective hack, and has been recommended by health experts including the Glucose Goddess Jessie Inchauspé and Andrew Huberman, a neuroscientist and podcaster. Just by changing the ORDER in which you eat your food you can have an impact on appetite. Eating carbohydrates first will spike your blood sugar (this is why restaurants serve bread first – it makes you hungry!). MUCH more than eating vegetables, or proteins/fats first. If you eat those items first instead, you'll blunt the impact of that glucose spike and achieve satiety much sooner. Plus, you'll have less of a sugar crash later on and be less likely to reach for snacks.

So, for example, if you have a plate with salmon, rice and broccoli on it (well done, very good), eat the salmon and the vegetables first, leaving the rice to last. The Glucose Goddess recommends at least some vegetables first, as this creates a 'mesh' in your gut to slow down the insulin spike. Another way to do this is to get some protein and fat in you before you sit down to eat. I would recommend everyone puts half a bloody avocado or an egg in their mouth before anything – and yes, that's BEFORE you sit in front of your kids eating because you *will* reach for a chip and pick pick pick.

If you have a sweet tooth, then have a protein shake instead of the ultraprocessed biscuits or chocolates – whack some fat in there like peanut butter (a pure one without any added vegetable oil) and a banana, too. People might say 'Oh, that smoothie is still high fat and high sugar!' and freak out, but it's more important to get you sated and weaned off these frigging processed foods. I know there's fruit sugars in bananas but those sugars are far, far better than the alternative: all the cereals and granola bars and the safety nets you've been addicted to and conditioned to think of as 'treats', when they're nothing of the sort. Try to remember that treats = cravings.

4. Eat nutrient-rich foods to trigger leptin

After everything you've read so far in this chapter, it goes without saying that you should ditch processed food from your diet as much as you can. If it doesn't go without saying, I'm saying it now! I've explained how much of a complete shitstorm processed foods are for everything from your gut microbiome to your appetite regulation, and how dangerous vegetable oils, emulsifiers and hidden sugars are for messing with your hormones. So, get yourself reading the ingredients list, or better, still, DON'T eat foods that come in a packet.

Instead, you should stack your diet with nutrient-rich foods that will hit your amino- and fatty-acid targets and ALSO trigger the 'I'm full' hormone response, i.e. our friend leptin. If you're sated with the food you eat, you'll also dampen down your ghrelin response, which as we know, tells your brain that you're hungry.

When it comes to the best foods to keep you fuller for longer, I'm not about to list loads of veg like salad leaves and broccoli.

They're good for you, of course, but they won't satisfy your nutrient targets. Nutrient-dense foods are those that contain the most complete proteins and nutrients your body needs to function optimally.

TEN BRILLIANT NUTRIENT-DENSE FOODS TO SATE YOUR HUNGER HORMONES

Eggs – you may have already guessed, but I love eggs in any form. They are a complete food, full of healthy fats, proteins, vitamins and minerals and *will* sate your appetite. They're also a fantastic natural source of choline (which I'll go into in more detail about in the *Why Does It I Feel Like I'm Losing My Mind?* chapter). I get my eggs from a local friend who keeps chickens, but if you don't have a handy local *The-Good-Life*-type mate, just buy organic ones from the supermarket.

Olives – in the same way that olive oil is one of the best cooking oils to use, olives themselves are really nutrient-dense. They're high in good mono-unsaturated fats, and are rich in plant compounds called polyphenols which have antioxidant properties. They're also fermented, which means they're great for our gut bacteria. Avoid olives in rapeseed and/or sunflower oil!

Oysters – all right, they're not always the easiest to get hold of, but oysters are packed with zinc, which works with leptin to regulate our appetite. They're also incredibly rich in omega-3 fatty acids, vitamin (HORMONE!) D, zinc and selenium, which are hard to get in other food sources.

Avocado – yes, they're a bit of a cliché these days, but avocados are amazingly nutrient-rich foods. Not only do they contain high levels of healthy fats (and so keep you fuller for longer), but they're also a source of vitamins C, E, K and B6, riboflavin, niacin, magnesium and potassium. They also provide good fibre (not All-Bran style fibre).

Fish – halibut is especially good for appetite-managing, as it has high levels of both protein and tryptophan. Salmon, mackerel and sardines are fantastic, too. All types of fish have great benefits, reducing inflammation throughout the body and allowing leptin to communicate effectively with the brain.

Ribeye steak – this fatty cut of beef contains lots of good fat and protein, plus plenty of B vitamins, zinc, selenium and some unique compounds only found in meat, such as carnosine, which supports healthy muscles as well as having anti-ageing benefits (who knew?).

Beef liver – even if you're a meat-eater, you might turn your nose up at the idea of eating liver. I get it. But there's a reason liver is called 'nature's multivitamin' – beef liver is packed with many of the hardest-to-get nutrients, as well as being a great source of protein. It contains heme iron, vitamins B6, B12 and A, plus loads of nucleic acid, which the body needs for digestion, muscle recovery, metabolic regulation and to strengthen your immune system. I add it to spag bol for the kids and so far they haven't noticed!

Greek yogurt (and other full-fat dairy) – Greek yogurt is especially nutrient-rich, as it contains lots of bio-available protein, calcium, magnesium, iodine and phosphorous. It also contains probiotics, which can support a healthy gut microbiome. Just make sure it's unsweetened! You can always add some stevia for sweetness.

Almonds – unsalted almonds are an amazing natural source of fibre, protein and good fats. What's more, they contain other vital nutrients, including fibre, vitamins E and K, folate, thiamine, magnesium, potassium and antioxidants. They'll fill you up far more than a bowl of Kettle Chips if you're feeling snacky. Remember to activate your nuts first by soaking them in salted water for ten hours.

Boiled potatoes – potatoes get a bad press, but in fact, they can be really good for managing your appetite. One study even showed that boiled potatoes made you feel seven times more sated than eating a croissant!* (those bloody croissants again). Even better, eat them cold after cooking – boiling potatoes and then cooling them increases the amount of resistant starch, which also makes you fuller for longer.

Collagen peptides – has an amino acid profile which the gut detects and stops grehlin in its tracks. I add a teaspoon of collagen to all my drinks throughout the day to help stop me snacking. Peptides are absorbed better than regular collagen due to the size of the molecules.

* https://www.researchgate.net/publication/15701207_A_Satiety_Index_of_common_foods

FAIL TO PREPARE, PREPARE TO FAIL

The lesson I want to get across in this chapter, is that you've got to get ahead of your cravings and ALWAYS PREPARE. What I mean by this is don't leave what you eat to chance. You've got to be a bit organised about things otherwise you will fall off the wagon. Like I've said, it's not because you're 'weak' – you're being constantly bombarded with subliminal advertising, temptation and negative-hormone-hacking habits! So don't blame yourself – instead, arm yourself so these food companies don't take control of your appetite – you do.

So, make sure you've got some decent proteins and fats around you to hand whether that's eggs, avocadoes or meat. Try out some of the supplements I've covered in this chapter. Move your eating window a little bit. Instead of ordering Friday-night takeaway pizza, invent a new ritual. Change your route to work so you don't walk past a Greggs – whatever you need to do to avoid your addictive-eating triggers and ghrelin-spiking release, plan it and do it!

THE CAT PLAN

To put all this into action, I want you to make a plan for a week where you will apply the **CAT** approach to what you eat – you'll **change** something, **add** something different and **try** something completely new for you.

Remember, we're all completely unique, and have very different eating triggers, so if something in particular appealed to you, like trying out berberine, or eating more eggs, then knock yourself out. But you might hate eggs, and instead have a massive fondness for sugar, in which case L-glutamine could be your friend. Or, you might be completely wired to eat a bowl of cereal the moment you get up and you're going to ditch that. Whatever it is, write down your three achievable CAT goals and try them for one week to see how it helps control your cravings.

This week, I'll change	e.g. delay breakfast by forty-five minutes
This week, I'll add	e.g. add an egg to my snacks (and eat it first)
This week, I'll try	e.g. a cold shower for twenty seconds after a hot bath

YOU CAN FEEL BETTER AND HAVE A HEALTHY APPETITE

If you find it hard at first, a good thing to remember is that if you've been eating loads of highly processed food, natural foods WILL taste bland and boring for a while. Your tastebuds will adapt back, though, and within only a couple of days. Natural food will start to taste good again, but you have to remove all the highly processed crap from your diet to feel the benefit. You've got to be a little bit patient, which is hard when we've all been programmed to get these dopamine highs from our food, suffer from the insulin crashes and constantly try and sate the ghrelin monster. But you'll get there; you'll end up feeling nourished, sated and without those horrific cravings. You can do it!

Don't worry if you give into temptation sometimes – we're

all human and I fall off the wagon all the time. If you've pressed the fuck-it button, the important thing is not to slide back into your old habits. You have all the information and protocols at your fingertips to get rid of cravings and manage your appetite hormones – now go out and do it.

DR E – HOW TO TALK TO YOUR DOCTOR ABOUT OVEREATING

We are increasingly seeing insulin resistance and pre-diabetes in our health screening of new patients. This is a lifestyle-related condition, and what's most interesting to me as a doctor is that we are finding this not only in the overweight population. It is present in people who are 'fit and healthy'. In our clinic we find the common element shared by those with insulin resistance is stress response, and if you are struggling with overeating my first suggestion would be to begin to take a look at what might be causing you stress.

The stress response state will:

a) raise your cortisol
b) raise blood sugar as your body thinks it needs to have energy available ready for an emergency
c) block insulin receptors from re-uptaking insulin, thus leading to insulin resistance.

Sprinkle sugar on this from your stress-induced diet of sugary snacks and then we see inflammation really kick off. Understanding what may switch on your internal stress response and making adjustments is, in my opinion, the foundation to blood sugar regulation.

Sugar dependency and cravings are all but unspoken of in conventional healthcare, and I'm so pleased Davinia is raising the connections between cravings and overall health as clearly as she has done above.

Why do medical professionals place so much importance on diabetes, or this state of inability to bring blood sugar down? Ultimately high blood sugar drives blood vessel changes which affect all organs, from kidneys to the skin and blood flow to your limbs and brain. Without the plumbing being healthy we can't deliver key nutrients to the cells of our organs and they are unable to function optimally. When organs break down it's a slippery slope.

Lets discuss the important symptoms to take seriously and attend your GP with:

Blood sugar dysregulation

- Being thirsty and having an urge to drink lots of water
- Passing urine a lot, weight loss, fatigue, recurrent infections
- Darkening of skin folds – typically the armpits and groin – suggests insulin resistance.

Treatment options include lifestyle support groups and oral and then injectable medication. Expert patient groups are an excellent source of support.

Weight gain

- Speak to your doctor about excluding underlying medical/hormonal conditions

- Review any medications that you may be taking on a regular basis
- Look at your lifestyle habits including recreational drug use
- Consider if there is a family history of weight-related medical complications – diabetes and coronary heart disease
- Review how weight gain may be affecting your quality of life, motivation, mood and confidence
- Be aware of associated risks from weight gain – lipid and triglyceride status, hypertension and insulin resistance, which can lead to metabolic syndrome

There are weight management services available from your doctor. Depending on your BMI (body mass index) there are various options available:

- Drug therapy – liraglutide and orlistat have been shown to be very helpful for the right patient
- Bariatric surgical intervention could be appropriate and considered

Eating disorders require support to help us think differently about ourselves and body shape, and ultimately our relationship with food. If you are concerned about your relationship to food, or someone else's, please do take this seriously and see a medical professional, as there is a lot of support available. Eating disorders can include undereating, overeating or binge eating. Warning signs of eating disorders include:

- Disproportionate concern about your body shape

- Emotional and mental health impact – stress, anxiety, low mood, social withdrawal
- Rapid weight loss
- Low BMI
- Menstrual disorders
- Binge eating (and shame after binges)
- Laxative use/ vomiting/ purging/ drug use after binge
- Over-valuing body shape
- Self harm

Medical testing:

These tests are not exhaustive, your doctor will be able to discuss with you which are worthwhile for your concerns.

BMI and examination including blood pressure

Thyroid baseline + adrenals baseline: TSH, T4, T3, morning DHEA and cortisol

Blood sugar response: HbA1c, fasting insulin and fasting glucose

Lipidology: HDL, Chol, LDL, triglycerides

Full blood count

Liver function test

Kidney function

Bone profile

FSH, oestrodial, progestoerone, testosterone

Integrative testing:

(suggested not exhaustive biomarkers – personalised by your practitioner after consultation)

Homocysteine, B12, folate – looking for methylation issues

Lipidology – ApoE genotype, apolipoproteins (breakdown of cholesterol), TG, vLDL, LDL, etc.

Metabolism – above + fasting insulin, adinopectin

Hormone tests:
Fatty Acid composition – omegas 6:3
MicroNutrient status: Mg, Zinc, Cu+, etc. – organic acid test is useful, or Genova NutrEval
Gut health: microbiome, leaky gut, parasites, funghi, etc. – comprehensive stool test
DUTCH Complete
(there are others, but these are the important ones to start with)

Tech:
Continuous blood glucose monitor – FreeStyle Libre 2 wearable patch to help you identify what foods spike your blood glucose
Ketometer – breath, helps you understand when you're in ketosis if intermittent fasting or following a low-carb, high-fat and protein diet
Health wearables such as Ōura use body temperature to track your cycle.

WHY DOES IT FEEL LIKE I'M LOSING MY MIND?

IN THIS CHAPTER, WE'LL FOCUS ON THESE HORMONES:

Dopamine: *our desire hormone, which also controls our focus and drive*
Serotonin: *our happy hormone that also affects memory and learning*
GABA: *which stops us feeling anxious*
Cortisol: *the stress regulator, which can go out of balance*
Acetylcholine: *our bodies' main neurotransmitter, behind memory and learning*

AND – WTF is the difference between a hormone and neurotransmitter anyway?

Ever feel as if you cannot remember the littlest thing? That everything is just *too much* and you can't deal? That you're completely overwhelmed, sometimes to the point where you're frozen in a state of permanent 'Nope, can't do this, forget it' and procrastination becomes your middle name? Hello! Nice to meet you, me too.

Honestly, I literally cannot believe sometimes how much I forget. Brain fog has really impacted my life; every time I walked into a room there would be about seven different subjects floating around my mind and I'd only be able to pay attention to the one that was shouting the loudest (do I need to pick up a pair of trainers for the kids or do I need to find a passport? I would have no idea). I lived in a constant state of fear that I had forgotten something important – and very often I had done.

Probably the worst (or best) example of my brain fog was when I took the entire family to Gatwick airport when we should have gone to Heathrow. I mean, I'd even *booked* Heathrow because it was quicker and cheaper on the train to get there, but instead I took us all to Gatwick only to be told there was no flight (of course). Oh, and by the way, I'm ten years sober by this point, so there's no chance it's hangover-related. I was so furious at myself.

HOW BRAIN FOG AND OVERWHELM MADE ME FEEL

People make jokes about brain fog and ditziness, *oh ha ha ha you'd forget your head if it wasn't screwed on*, all of that, but for me, it feels like a real place of pain. I felt especially shitty when someone gave me an AA coin when I was two years sober (it's a thing that you're given in recovery as a mark of how far you've come). He was such a nice man and kindly gave me this coin he'd had for years – but within a couple of days I'd lost it.

He was so decent about it, saying, 'Don't worry, my wife's exactly the same, I know you're going through a lot of self-hatred and pain about this.' And he was right, it felt SO painful. I was really ashamed by the fact that I'd lost it, and it just fed into the feeling I had that I was a 'bad' person who couldn't organise myself properly. Now I know that this was probably down to my ADHD, which I'll explain in a bit more detail later on.

Getting that hunted, panicked feeling is something that loads of us deal with, and I notice mine's more prevalent in the early days of my cycle. I'm incredibly jumpy, and can actually *feel* my mood drop from my head down to my stomach (you know that hot wave feeling when you get bad news? Just like that). At these times, I just can't seem to tolerate the day-to-day stuff that I usually can; I'm hypersensitive to the kids shouting and am completely consumed with dread that something bad will happen to my family. Most of the time I can put things into perspective, whereas on these days I'll wake up in the middle of the night convinced that my youngest son, Jude, will get run over by a car on his bike (this wasn't helped by the fact he *did* ride out across the main road once, but when I'm feeling like this I can't let it go).

A HIDDEN EPIDEMIC AFFECTING US

When I posted on Instagram about brain fog, and the overwhelm it brings with it, I couldn't believe the number of responses I got from women who've been struggling with it. Brain fog is a huge issue and affects so many of us in multiple ways. Jo wrote, 'It's like I'm in a dark room and have no idea where the light switch is', whereas Suzie said, 'You just feel really stupid, like you're quietly going mad. You know there's something wrong but it's so subtle you don't know what. It's like your own body is gaslighting you.'

It's so detrimental that it can negatively affect our jobs and relationships, too. Jennifer wrote, 'I am in a job [where] mistakes can have a big impact on people and these days I have to double, triple check everything so I don't come unstuck. I can't even spit my words out regularly so leading meetings can look unprofessional like I don't know what I am doing.' Midwife Amanda Jane added, 'I remember driving to work on an early shift and feeling like that expression "the lights are on but no one's home" – I didn't feel like my mind and my body were connected at all . . . I'd forget the names of things or the words I needed and that was very frustrating . . . I felt like I was losing my marbles!' Sue says, 'There are days when I can't get my head round tasks at work that I've done a thousand times . . . it feels like I can't think straight, and everything is going in slow motion.'

Feeling like you're losing your mind is such a horrible, unsafe place to be emotionally. Whether that's brain fog, feeling like you simply cannot focus, procrastinating like it's an Olympic sport or feeling overwhelmed by life, it's REAL, and can make us feel less-than, like Natasha posted on Instagram: 'I feel like

I've lost me! I used to be a really self-confident city girl. Now I'm either angry, sad or mad.'

For so long, though, these emotions have been completely dismissed, because they're hormone related and until recently, haven't been well understood. A lot of these symptoms are also linked to perimenopause and menopause, which is why I'd recommend reading the *WTF Is Up With My Hormonal Cycle* chapter, too. But thankfully, there is now loads of research that's been done on the hormones that cause these symptoms, and this means that there's lots we can do to help ourselves feel better.

THE HORMONES THAT AFFECT THIS ARE . . . BUT FIRST . . .

. . . WTF is the difference between hormones and neurotransmitters?

Before we get into a deep dive on the hormones behind all these effects, I wanted to explain a bit about neurotransmitters vs hormones. There's a good reason for this – I don't just want to give you a science lecture!

I thought this chapter was the perfect place to explain the difference, as funnily enough, the main neurotransmitters – serotonin, dopamine, GABA and acetylcholine – are massively relevant for brain fog, focus and overwhelm. I'll try and keep it simple, with a few quick facts:

They have different effects

Neurotransmitters are brain signals that impact our thoughts, feelings and automatic responses like movement and heartbeat. When neurotransmitters are severely imbalanced, people can develop psychological mood disorders like depression, anxiety and insomnia. Hormones are chemical signals that can massively affect our mood too, but also look after a lot more functions, such as growth, development and reproduction.

They're part of different systems

Hormones are produced by the endocrine system, a network of glands and organs across our body. Neurotransmitters are part of the nervous system, our body's 'command centre', which travels from our brain, along our spinal cord to different parts of the body.

They travel through our bodies in different ways

Hormones are transmitted through our bloodstream, and usually travel some distance from one place (a gland) to their point of action (another gland or organ). Neurotransmitters travel a tiny distance across the 'synaptic cleft', which is a space that separates two neurons, mainly in our brains.

Neurotransmitters and hormones work at different speeds

Neurotransmitters take literally milliseconds for us to feel their effect, like that immediate dopamine hit after you get a 'like' on social media. With hormones, the effect can take anything from a few seconds up to a few days in some cases.

Some molecules can be both a hormone and a neurotransmitter at the same time

Look, I'm sorry about this. After explaining the differences between a hormone and a neurotransmitter, the last thing I want to do is confuse things. But it's not my fault, OK? (Sometimes science doesn't make sense, because these people are maniacs and just want to confuse us).

Basically, some molecules can be *both* a hormone *and* a neurotransmitter, for example, serotonin, adrenaline and dopamine. They're involved in so many different mechanisms in our bodies that they function as both. So, when serotonin is released, it works as a neurotransmitter that sends signals via our nervous system in the gut, and is *also* released into our bloodstream where it works as a hormone.*

So, there you go – now you know the difference and you can impress your friends at your next party with this amazing knowledge (on second thoughts, maybe *don't*). So let's look at the way these neurotransmitters-and-hormones work on our focus, brain fog and feelings of overwhelm.

Serotonin

Famous as our main happy hormone, serotonin makes us feel lovely, safe and cosy. It's released by the pineal gland in the brain, but the vast majority of our serotonin is manufactured in the gut, starting off as tryptophan, an amino acid. This then is converted to 5-HTP, which then becomes serotonin. This is

* https://atlasbiomed.com/blog/serotonin-and-other-happy-molecules-made-by-gut-bacteria/

why having a healthy gut microbiome is massively important for supporting our serotonin levels.

In order for our brain to feel the benefits of serotonin, i.e. cross the blood–brain barrier, you must have adequate amounts of vitamin D and ideally the MK-7 form of vitamin K2 (which is more bioavailable and lasts longer in the body).

Serotonin has physiological effects on SO MANY functions (as you'll see going through this book), but for this chapter, it's really relevant to mood, memory and learning. If your serotonin levels are low, it's likely you'll be struggling with these issues, as well as many more.

A side note on serotonin and poo

Now, this is interesting, because who doesn't like a quick fact about poo? Serotonin also regulates our intestinal movements, so without having a hormonal test, you can get a quick gauge of your serotonin levels this way, too. Constipation can indicate low serotonin levels, whereas diarrhoea may mean they're too high. Brilliant, right?

GABA

If you've already read the Sleep chapter, you'll already be familiar with our friend GABA (gamma-aminobutyric acid. Honestly, don't even *try* to say it). GABA is a fantastic, anxiety-inhibiting neurotransmitter that slows down our whirring brains and promotes a calm, relaxed feeling.

It's produced in the brain, and its main function is to regulate our immune response, controlling our fear and anxiety when our neurons become overexcited. So if you're a nervous wreck or struggling with feelings of guilt and OCD, you're probably

GABA-deficient. (Being low in GABA will also negatively affect our sleep patterns.)

We can't buy GABA in supplement form in the UK (though you can have it prescribed privately), so we have to hack it in other ways – mainly from foods and exercise.

Acetylcholine

You might not have heard of this before, but it's actually the most common neurotransmitter in our entire bodies! It's based in the central nervous system and is in charge of muscle control, memory and sensations. Because it regulates our brain speed, if we have low acetylcholine levels we might have memory problems, difficulties in learning and thinking creatively.

We can buy choline in supplement form, which I take in the mornings, or we can get it from certain foods – especially eggs (more details later on). It's vital for helping with brain fog, thinking more clearly and improving your memory.

Cortisol

Cortisol is one of those hormones that gets a bad reputation, but it's essential for life – we just need to harness it at the right time! Known as the stress hormone, it's released by the adrenal glands once every twenty-four hours in a big cortisol 'pulse', which gets us out of bed, alert and ready for the day. It's at its peak about thirty minutes after we get up.

It's a motivating, movement hormone, but at the wrong time or at the wrong levels, it can leave us feeling anxious, on edge and hunted. Cortisol dysregulation is more and more common in our hyper-driven modern lives and can be triggered by everything from too much stress, to overuse of our phones to watching thrill-seeking movies late at night. We want to manage our cortisol so we benefit from it during the morning and not spike it throughout the day and evening.

If your cortisol is high, two supplements that can really help are ashwagandha and rhodiola.

A super-quick side note on insulin

I'm not going to go into massive detail in this section, because it was covered earlier, but it's worth remembering that a dip or peak in your blood sugar will affect mental alertness too. For advice on how to bring your blood sugar into balance, plus why leaky gut can cause issues like brain fog and memory loss, have a read of the *Why Can't I Stop Eating?* chapter.

Dopamine

Dopamine is the hormone I like to call the double-edged sword. It's known as the desire hormone, but on the flipside of that is pain, also known as craving. Dopamine is behind all those feelings, as it drives our motivation and desire to do, well, ANYTHING outside of ourselves! Our dopamine levels influence our mood, attention, motivation and movement, and very importantly, our reward system. If you read the *Why Can't I Stop Eating?* chapter, you'll remember the Croissant Pathway – and how important dopamine is for driving our behaviours.

Like I've explained before, dopamine can make us do fantastic things – like run a marathon, pass exams, or give an amazing speech – but it can also tip us into addiction. Your body has a constant baseline level of dopamine, and if yours is low (like mine), you're likely to have a huge predisposition for activities that will give you a dopamine boost.

Those activities could be wholesome ones like skydiving and running, or they could be way more damaging habits like gambling, alcohol, relationships or even shopping. The problem with trying to increase your dopamine through addictive habits or substances is that they have a really short-lived effect and can actually *decrease* your baseline level. This is why addicts end up seeking more and more of whatever they're relying on – whether that's food, drugs or alcohol – to achieve the same dopamine feeling.

Whether or not you have addictive tendencies yourself, we're all still at the mercy of our baseline dopamine levels. Andrew Huberman, the neuroscientist behind one of my favourite podcasts, *The Huberman Lab*, describes dopamine brilliantly. His explanation is that dopamine is our 'currency' because it's how we track our experience of pleasure. It's not just as simple as getting a dopamine hit from something – it's dependent on our baseline levels and also what we did leading up to that experience – basically put, everything is relative, and dopamine is too!

How dopamine affects our focus and drive

The bad news is, if we repeatedly do the same dopamine-boosting activity over and over again, our threshold for experiencing that pleasure goes up and it doesn't have the same effect. I'll explore how we can hack this later on, but it means that the solution ISN'T doing the same thing repeatedly.

Here's an example – you're at your baseline dopamine level, and then you peak it by having a creme egg. But, as you take the last bite of chocolate, you dip below the dopamine baseline.

Instead of feeling happy and satisfied, you feel a craving for something else to replace that missing one – e.g. another creme egg or a mint Aero (my kryptonite). That 'want' is hardwired in, because once that experience is over, you've triggered the pain mechanism, and it feels horrific.

If you're pushing yourself into dopamine highs all the time, you'll end up pushing yourself below baseline levels and then you'll do *anything* to get out of that low. It's a brain thing, not terrible willpower. Your body is telling you to go and get that dopamine high again to reach the baseline level.

That's why drug addicts will rob a bank to get the money to get their next fix. It's why people who are morbidly obese cannot stop ordering another McDonald's. They want to escape the pain of the craving. On the flipside, you have people like Susan Wojcicki, YouTube CEO, or Spanx founder Sara Blakely, who are these hyper-driven entrepreneurs, and repeatedly seek out business success as their dopamine hit – so many of these hugely successful people are definitely low-dopamined! But this behaviour can just as easily tip into addiction.

We need to remember that our cravings are about biochemistry *not* willpower. So, if you're struggling with motivation, attention, and a massive dose of 'meh', then you're dealing with low dopamine – either through having a generally low baseline, or because you're doing too many dopamine-spiking activities (Insta or Facebook scrolling, anyone?). Instead, you can manage your dopamine positively, to keep your baseline at a healthy level and increase motivation and focus. I'll show you how to do this further on in this chapter.

My ADHD diagnosis – and why it's linked to dopamine

I'd always thought that I was just a bit of a scatterbrain – after all, that's what I'd been told since childhood. At school, I was known as a real dolly daydream and I was a complete scruff, always missing a tie or a sock, and never had a pen to hand. I was constantly being told, 'You need to organise yourself better,' and when I wasn't able to, I felt so bad about it. I would wonder *why* I couldn't sort myself out, and why trying my best wasn't good enough for most people (even now, I feel a bit ashamed about my lack of organisation).

I went through my whole adult life, with all my addictions and issues, before I understood I had ADHD. It wasn't until I was filling in a form for one of my sons, who has ADHD, a couple of years ago that I suddenly thought, 'God, this is me too.' It led me to get my own formal diagnosis, which was such a huge relief, I cannot even tell you.

Straightaway, I took a deep dive into what's behind it, and found that there is a strong link between low baseline dopamine and ADHD (as well as a lot more, of course, like genetics). It made realise, 'Ah, maybe THAT'S the reason I drank so much!' Most people drink alcohol to relax; I drank it to get things done. It made me active and able to <u>do</u> things, like pay bills, tidy up, be organised – all the stuff I couldn't manage before. I was hacking my hormones detrimentally by boosting my baseline dopamine with alcohol, and obviously, as you know, that led to all sorts of problems.

Now, even though I still struggle with low dopamine levels and general get-up-and-go, I hack my natural biochemistry in less destructive ways – for me that looks like running, house music and caffeine. Now that I run my WillPowders business, too, I have people who work for me, which means I can delegate some of the tasks that would have floored me before. My life is full of spinning plates, and trying to organise everything had become a constant nightmare of white noise to my ADHD brain. Things aren't perfect now – far from it – but understanding what's behind my brain, my overwhelm and constantly jumping from one project to the next, has been so helpful.

HOW THESE HORMONES ALL WORK TOGETHER

You might not have ADHD, but we can all experience the fallout that happens when our hormones are out of whack. I would say, though, if any of the above section resonated with you, I'd recommend seeking further advice from a medical professional about a potential ADHD diagnosis. Historically, women with ADHD have been *massively* underdiagnosed because we generally display more of an ADD 'dolly-daydream/scatterbrain' type diversity and are less 'hyper', but thankfully things are changing on this front now. I'm not on any typical ADHD medication, but I'm taking LDN (low-dose Naltrexone) which seems to be helping my symptoms. It's a new kid on the block, so ask your doctor to do a deep-dive. So far I've had no side-effects like anxiety.

Like I've said before, these hormones do NOT work in little individual silos! They are all doing a dance together, so if one is low, or too high, that will have a cascade effect on others. If your cortisol is dysregulated, you'll feel hunted and stressed, and will probably have lower levels of GABA, which is linked to anxiety. In turn, this will mean that your serotonin is low. If you're low on dopamine, you'll also be low on noradrenaline and adrenaline, as the body uses dopamine to create them. It just goes on!

Safe to say, when these hormones aren't working at their optimal levels, they can affect your memory function, motivation, brain fog – all those traits that can lead you to feel that yep, you are in fact losing your mind. But we don't want you to feel like that – and you don't need to. We can hack these complicated and complex hormones with food, supplements, exercise, and a few cheeky little free tricks.

SIX EASY WAYS TO FIND FOCUS AND CLARITY

1. Overcome overwhelm with salts

Right – this first part is all about managing your cortisol levels. Look, I know you've got brain fog and maybe I'm asking too much of you when you're already overwhelmed – sorry! – but you could also read the chapter on sleep to get a good insight into how to regulate your cortisol release with morning activities. Remember, we don't want to peak your cortisol or reduce it, we just want it to be released at the right time – in the morning – not spiking throughout the day where it could tip over into panicky stress.

A great way to boost cortisol in the a.m. is to kick-start your adrenal response with salts. Sea salt has got so many minerals and electrolytes in it that are key for every cellular function in the body. I know salt (a bit like fat) has been demonised by doctors, but a natural unprocessed salt like Maldon can be really beneficial. Because cortisol regulates our sodium (salt) levels, having some in the morning is a great way to get your adrenal glands working healthily at the right time.

Some biohackers recommend drinking some water with half a teaspoon of salt and then lying back with your legs up on the wall. Nice idea in theory, but this just *wouldn't work* for me at all. I've got three small lunatics shouting at me in the morning, I don't have time to lie there waiting for my adrenal glands to wake up! (The world of biohacking can be very blokey, and honestly, some of the things they recommend just can't be done when you've got kids and commitments – I get it.)

Instead, I get around this by mixing a decent electrolyte powder with water at the start of the week in a big jug. I keep it in the fridge and pour myself a glass of it first thing, so it does its work and I don't need to worry about it.

2. Tackle your triggers with white noise

Another really interesting fact is that white noise can also help increase our focus. White noise is a type of noise that includes all sound frequencies the human ear can hear, played at a similar intensity – so things like, whirring fans, humming air conditioners, and even traffic noises (if they're not too loud). Think of the kind of low-level buzz sound that feels restful rather than annoying.

If you've had a baby, you've probably used white noise to settle them down – either by switching on one of those white-noise teddy bears or by putting them in the back seat of the car and going for a drive! It's amazing the calming effect the sound has on them, and it works for us adults, too. I often fall asleep with the telly on, because Matthew likes to watch daft documentaries about digging for gold in the Congo, or whatever. I'm completely unengaged with these, so they can help me relax and drift off. The important thing is I'm not listening to the actual sounds themselves, it's just background noise.

Studies have shown that white noise can not only improve our concentration and memory, but it can activate our dopamine pathways, too.* You can either introduce natural white noise with the washing machine, fans, sounds of nature from your garden (if you have one), or even just play a white noise video loop on your phone.

Intentionally introducing a bit of white noise into your daily life – whether that's before you get out of bed, or while you're working – might help counteract some of that fuzzy sense of overwhelm and inability to get anything done (as well as managing that stress when the dogs have peed all over the couch – or maybe that's just my house). Also, there are green, blue and yellow noise apps available. You don't have to have an ADHD diagnosis to benefit from subliminal sounds.

3. Fix your brain fog with good fats

One of the mantras I say over and over again is that FAT IS A FUEL – it is actually good for you to stack your diet with clean fats like MCT and olive oils. As humans, we've always had fat in our diet, and it's only been during the last fifty years that this mania for low fat has taken over (and where has it brought us? Worldwide obesity, dementia and type 2 diabetes crises! Just saying).

The reason we need fats is that they are our system's building blocks: helping us transport nutrients around our body to

* https://whisbear.com/en/blog/how-to-use-white-noise-safely/

actually *make* the essential hormones and neurotransmitters that will make us feel happy, content and focused. Without the right sorts of fats in our diets, our glands simply CAN'T produce hormones and neurotransmitters properly and at the right levels.

It's been known for a long while now how beneficial the MCTs (medium chain triglycerides) found in coconut oil are, because they bypass your normal digestive system and go straight to your liver where they're converted to ketones. This might be going a bit science-y (bear with me, here) but essentially, these then are involved in the production of ATP, our cell's 'petrol'.

ATP stands for adenosine 5'-triphosphate (bloody hell), but what you really need to know is that it improves alertness, memory and mood, which is just what we all need when we're struggling with brain fog. Research also shows that the MCT in coconut oils increases antioxidant levels in the brain and also serotonin, which has an anti-stress effect.

Every morning I have MCT keto powder in my coffee, and it's had a transformative effect on my mood and focus. It boosts my noradrenaline (adrenaline in the brain) because it crosses the blood–brain barrier and gets right in there. Some days I can literally feel a shift in my brain mechanism and almost feel a tiny weeny bit drunk from it, which for me, is actually my preferred state. (MCT oil seems to be the closest thing I can get to a pinot grigio these days, which is fine by me!)

It's interesting to see, too, how different people experience it; a friend of mine says she can feel a teaspoon of it hit her straightaway. To her it's like the feeling of a glass of champagne, making her a little bit bubbly and optimistic.

4. Support serotonin and GABA with a strong microbiome

Our guts produce 95 per cent of our serotonin, so it is absolutely essential we support this with developing a healthy gut microbiome. If you've already read the *Why Can't I Stop Eating?* chapter, you'll have already seen why processed foods are an absolute shitstorm for our microbiome – not only do they make us overweight and addicted to junk, they also inhibit our production of these hormones that make us feel calm, cosy and safe.

We need to take our head out of the advertising propaganda that equates eating crap with being a 'treat', and instead give our gut the best environment to produce serotonin and GABA by stacking our diets with fermented foods. They've been proven in scientific studies to increase the diversity of our gut microbiome, which in turn allows it to produce the hundreds of neurochemicals that influence our mood and behaviours.

Check out fermented foods like kefir and raw cheese if you like dairy, sauerkraut or kimchi if you like food with a kick, or kombucha if you're looking to swap your fizzy drink habit for something less damaging to your health.

Any alcohol, just like hand sanitiser, kills bacteria both good and bad; so before and after drinking alcohol make sure you load up with pre- and pro-biotics to support your mood (hangxiety anyone?!).

5. Eat to support focus and clarity

There are so many different foods that help us produce the hormones and neurotransmitters that we need for focus, drive and motivation (and none of them are salt and vinegar Square crisps, sadly). I could write down lists and lists of what these individual items are, but that would just be boring, wouldn't it? Instead, here are a few food and drink ideas for super-charging your hormones for focus.

Key: G = GABA-boosting; S = serotonin-boosting; A = acetylcholine-boosting

Drinks
Tea (G)
Kombucha (G, S)
Hot chocolate (whizz up cacao powder, dates and MCT keto powder, bovine collagen peptides and/or full-fat milk) (S)

Meals
Poached eggs (A, S) on sourdough toast (G, S)
Omelette (A, S) with mushrooms (G, S) and/or cheese (S)
Halibut (G) or salmon (A) with rice (G) and spinach (G) or broccoli (G, S, A)
Grilled tinned mackerel (G) in tomato (G) sauce (not the ones with sunflower oil!) with melted cheese (S) on sourdough (G, S)
Turkey or chicken (S) with broccoli (G, S, A) and brussels sprouts (A). I generally cook all my meat in butter, garlic and herbs.

Snacks
Nuts and seeds (G, S)

Bone broth protein shake (G, S, A)
Hard boiled egg and spinach (G, S)
Full-fat yoghurts and fermented foods (G, S, A)
Grass fed beef butty with sourdough bread (G, S, A)
Cheese butty (S)
Home-made soups with a bone broth base and bovine collagen peptides. (G, S, A)

6. Create calm and contentment with connection

I absolutely LOVE this hack, because it's so straightforward and accessible. You can actually increase your serotonin levels by finding connections within your community. This isn't just woo-woo hippy stuff, this is actually how recovery programmes like AA work – being part of a group with shared interests/ intentions actually boosts serotonin production and so you get those lovely comforting vibes that help support you (and stop you reaching for a glass of wine).

In this case, boosting your serotonin through better connections with others will help counteract that hunted, panicky feeling, and improve your mood and memory function. Don't worry – I'm not going to ask you to join a book group or something – honestly, I couldn't think of anything *worse* than being told to add something else to my frigging diary.

You can find connection anywhere, even at the local supermarket – it's a great reminder of how we are not alone. When I feel disconnected, I know that I just need to find that moment of connection with another woman. I spend a lot of time at home, because that's where I work, there's the kids and the dogs, then there's me and Matthew during the daytime and

I love him, but *he does not get it.* So, if I can't meet up with a friend or colleague, it's so helpful to take myself down to Tesco, and just look around at every single woman and know that we're all having a shocking time of it on some level. There's some real reassurance in that.

If you've got kids, even seeing a small child acting up in the supermarket can be a moment of connection. There's that smile you share with another woman – not a smile of 'Oh I'm so happy your kid is being a maniac' – but more of a nod of recognition. You give that smile of familiarity to the mum, roll your eyes together, and think, 'Bloody hell, it's tough for all of us.' Not a word has been said out loud, but you've made a connection with another human, and that's a mood booster. So go out and find connection, wherever you can.

EASY DOPAMINE HACKS – PLUS THE ESSENTIAL WAY TO MAKE THEM WORK

Increase your drive with caffeine

Now, if there's ever been a biohack that will go down well with most of us, this has got to be it: caffeine can actually help you boost your dopamine levels! Brilliant, isn't it? Caffeine doesn't actually increase dopamine itself, instead, it works by opening up the dopamine receptors D2 and D3, which means you're better able to feel more of dopamine's effects.

I'm not advising you to mainline Starbucks all day long, but I want to reassure you that it's not a bad thing to have your coffee or tea in the morning (with MCT oil or powder, of course). It's the reason why I use caffeine myself, tempered with L-theanine to take the edge off. I find caffeine really boosts my mood and focus, whereas without it the day just feels like a real slog.

I know that in an ideal world, I'd be so balanced that I'd be 100 per cent focused and organised without caffeine, but right now, what can I do? I have kids and a business and I need to be up at six every morning. Maybe in the future I'll be able to function on some flipping ylang ylang tea or whatever, but right now, needs must. And that's what biohacking is all about – working with where you're at, rather than aiming for perfection. For me, knowing that a bit of caffeine helps my dopamine levels makes it even better.

Exercise for a mood boost

Exercise is absolutely amazing for giving you a dopamine (and serotonin) boost, and it's one of the main ways I regulate my naturally low levels. It's not fully understood exactly why, yet, but regular exercise has been shown to increase levels of dopamine in the blood,* and over time, it actually rewires our reward system, making our dopamine receptors work better, too.† I love running, as I find it suits my driven nature best, but that might not work for you. It really doesn't matter *what* exercise it is – swimming, cycling, yoga, tennis, even sitting in a sauna – literally anything that gets your heart rate pumping and your body moving will trigger dopamine and endorphin release (another feel-good hormone), as well as supporting GABA, which reduces anxiety.

* https://www.livestrong.com/article/251785-exercise-and-its-effects-on-serotonin-dopamine-levels/

† https://greatergood.berkeley.edu/article/item/five_surprising_ways_exercise_changes_your_brain

I completely understand how hard it is to get motivated to exercise, especially if it's not already a part of your routine. You can start slowly and build up – I certainly wasn't brilliant when I started running. It's completely fine to stop and start between walking and running, if that's what you're trying out, but don't push yourself too hard. Doing too much high-intensity exercise that puts excessive pressure on your adrenal glands could actually trigger excess cortisol, which is what you DON'T want. You should feel good after exercising, not horrific or exhausted, so don't overtrain, and ensure you have rest days in between exercise sessions. If you absolutely can't get motivated to move, try some Wim Hof breathwork for a week. It will improve cardiovascular health while lying down.

Get into cold water for alertness

Yes, for every 'have a coffee!' easy biohack, there's one which isn't so simple to do, and that's our good old friend COLD WATER EXPOSURE. There's no getting away from it, cold water is an incredibly beneficial tool to hack our hormones positively in all sorts of ways. If you've read the *Why Can't I Sleep?* chapter, you'll have read how cold showers can help stimulate sleep-supporting hormones.

Not only this, cold water also has amazing effects on clarity and focus, too – studies show that cold exposure increases levels of noradrenaline and dopamine in the bloodstream by as much as 530 per cent and 250 per cent respectively.* I mean, those are insane increases, right? Even better, the increase in dopamine from cold exposure was shown to be a sustained

* https://link.springer.com/article/10.1007/s004210050065

increase, and didn't result in the baseline crash that comes from other dopaminergic activities. You may not *want* to stand under a cold shower (who does) but you can't argue with the evidence.

So, build up to it slowly – either by gently turning your shower dial down from warm to cool (don't just jump into a freezing shower, you'll jump right out again), or by heating yourself up in the bath so much that you end up craving that feeling of cool water on your skin. I love ramping myself up in a sauna pod until I'm blistering hot, and then a cold shower is exactly what I want.

Heat helps

Poor liver function reduces dopamine synthesis, so be sure to support your detox pathways after a night out. I'm a huge fan of infra-red saunas, which help the liver detox and have also been associated with the natural anti-depressants in our brain: dopamine, norepinephrine and serotonin. These also help to lower the level of cortisol which is associated with stress and tension. A brilliant detox is the niacin sauna protocol (see my *It's Not a Diet* book for details).

But DON'T fall into regular patterns

However – with dopamine you've got to be careful. It's not simply a matter of just following the advice every single day – timing is everything. And the advice I'm going to give you might seem bonkers, but whatever your chosen dopamine booster, DON'T DO IT ALL THE TIME. Whatever it is, you need to mix things up and keep it random.

This is because if you do the same habit over and over

again, your dopamine response ebbs away and loses its power. Remember what I was saying earlier about addicts and lowering their baseline dopamine response? The reason a drug addict needs more and more to achieve the same dopamine-fuelled 'reward' (and same with a food/alcohol/exercise addict!) is because they are constantly spiking their dopamine levels, which ends up depleting their baseline. Remember that we only have a certain amount of dopamine available to us – it's not an infinite source that we can keep pumping out over and over again forever. If we do, eventually, we will feel less pleasure from the same thing. Waaaaaaah.

Your body gets into patterns and learned responses so we NEED spontaneity in order to trick our hormones, that means our dopamine-promoting activities have to be intermittent, random and surprising. Think about it – the reason we find dopaminergic activities like social media so compelling is because we literally *don't know what we'll find* as we carry on scrolling downwards forever.

It's also why gambling is so addictive too – there is no certainty in it, and our hormones love that, so we get those dopamine spikes from not knowing what will happen from one moment to the next. This is one of the main reasons why gambling is one of the most lucrative and successful businesses on the planet. The house *always* wins.

How to keep it random by flicking a coin

So how do we get around this? It's actually quite simple – by the flick of a coin. By all means add in lots of healthy dopamine-promoting activities, but rather than rigidly scheduling each one, decide whether to do it each day with a flick of a coin. Make it an easy yes/no situation; 'Shall I go into the cold today or not?' 'Shall I run with my headphones on or not?' 'Shall I use white noise for focus today or not?' Flick your coin and there's your decision.

We're creatures of habit, programmed to fall into routines and I'm no exception. I have a tendency to stack my dopamine in the mornings by drinking a coffee while listening to my favourite 90s house music playlist on my headphones to ramp me up to go on a run. It's so easy to want to do that every time, because I get such a hit from it – but if I keep doing it day in day out, it loses its power and starts to feel a bit beige. If I don't do it for a while, and just run with the sounds of nature as my soundtrack, when I *do* go back to my playlist, I'll get goosebumps again. It's amazing, and shows the power of keeping things a little unpredictable as far as your body is concerned.

So, in any way you can, mix it up. Flick that coin. We have to keep things exciting, and part of that is unpredictability. Our dopamine response is so essential for giving us those feelings of motivation and drive, so we need to look after it carefully, and not wring it dry by overdoing these dopamine-spiking activities.

Some supplements to help clarity and focus

As I've said before, balancing our hormones isn't a one-size-fits-all game. We all have our own personal, optimal balances that are completely different, and just taking loads of supplements won't necessarily solve all our issues. There isn't a standard level of supplementation that works for everyone, so you have to biohack smartly and use trial and error to find what works for you as an individual. These are a few ideas for supplements that may help you. As ever, try one at a time and see how you get on.

Acetylcholine

As we've seen, this is our body's main neurotransmitter and helps us learn. We get choline from certain foods (liver, eggs, broccoli, fish, etc.), but we can also boost it with a choline supplement. A daily dose can be 100–500mg – I take my tablet in the morning. One egg delivers 147mg – the perfect kids' breakfast.

Mucuna pruriens

This is a natural adaptogen that has its basis in Ayurvedic medicine. It has high levels of a dopamine pre-cursor called L-dopa, so supports the body's production of this fantastic hormone. A daily dose can be 15–30mg.

Nicotine

I know we all think of nicotine as completely evil, but it's actually a nootropic (brain-booster) in micro doses. I'm not suggesting you spark up a Marlboro Light, but you can put a nicotine patch on and boost your mood and focus (make sure it's a small dose, around 2mg). Of course, nicotine comes with addictive properties, so be careful!

EPA

EPA is a type of omega-3 fatty acid that's found in fish oil. It's amazing for helping our brains function well so is a popular option if you're suffering with brain fog. A good place to start is with a daily dose of 1,000–2,000mg. I've found it's superb for ADHD and also for kids.

Valerian root

As we can't get GABA supplements in the UK (grr), we can boost it with other supplements, one of which is valerian root. This natural extract has been found to increase the levels of GABA in the brain, lessening anxiety. As it's also used as a herbal remedy for sleep, you should take a smaller dose – around 120–200mg three times a day – to avoid feeling sleepy.*

L-theanine

Can cross the blood–brain barrier to promote relaxation and improve GABA levels.

GPC

Glycerophosphocholine is a precursor to the neurotransmitter Acetylcholine. Found in liquid form, this is a fast-acting solution for short-term focus support.

Nootropics/smart drugs

Nootropics is a term coined by Silicon Valley coders as a name for compounds or supplements that enhance or improve cognitive performance. These include N-acetyl L-tyrosine, L-phenylalanine, nicotinamide (precursor to NAD), and glucuronolactone to name a few. Consider implementing these when you need motivation to move or do paperwork!

YOUR WEEKLY PLAN FOR FOCUS

To give yourself the best chance of fixing that brain fog, procrastination and horrible hunted feeling, try out a few things I've suggested in this chapter. Because there are a good few

* https://www.healthline.com/health/food-nutrition/valerian-root#dosage-for-anxiety

hormones at play here, to start off with you could pick one thing for each hormone, try it for a week and track how it helps or not. Like I've said before, we are all wired completely differently, so what works for one of us won't work for another. With a good combination of some of the hacks below, you'll feel so much better.

Serotonin supporters

- Connection and community
- Fermented foods
- MCT oil
- Good fats (for more info, read the *Why Can't I Stop Eating?* chapter)
- EPA supplement

This week, I'll try________________________________
This is how I'll do it______________________________
Result____

GABA boosters

- Tea
- Fermented foods
- Exercise
- Foods rich in glutamic acid (bananas, brown rice, fish, etc.)
- Valerian root supplement

This week, I'll try________________________________
This is how I'll do it______________________________
Result____

Acetylcholine enhancers

- Eggs
- Fatty meat and offal
- Cruciferous vegetables (broccoli, cauliflower)
- Dairy
- Choline supplement

This week, I'll try______________________________
This is how I'll do it___________________________
Result______________________________________

Cortisol calmers

- White noise
- Morning light (see the Sleep chapter for more info on this)
- Sea salts

This week, I'll try______________________________
This is how I'll do it___________________________
Result______________________________________

Dopamine drivers – but remember to flick the coin!

- Caffeine
- White noise
- Exercise
- Cold water

This week, I'll try______________________________
This is how I'll do it___________________________
Result______________________________________

DR E – HOW TO TALK TO YOUR DOCTOR ABOUT BRAIN FOG AND OVERWHELM

Brain fog is commonly used as a term to describe mild forgetfulness, lack of focus and clarity. When women complain of 'brain fog' I find it is often related to one of four causes: the hormonal changes associated with pregnancy or postpartum; perimenopause; a stressful period of time leading to adrenal and thyroid gland imbalance; long Covid symptoms.

In our clinic we have found that hormonal imbalances that affect our cognition need addressing in four interrelated systems.

- Gut: digestion and absorption. Even if you are eating well, if you're not absorbing your food then you may not be well nourished.
- Inflammation: this is the process of an overactive immune system driving a reaction that can be manifesting in a multitude of ways, from rashes to hives, joint aches and pains.
- Micronutrients: key nutrients that are necessary for your body to function optimally, such as vitamins B, C and D and minerals such as copper, zinc and magnesium
- Hormones: hypothalamus – pituitary – adrenal – thyroid – sex hormones (HPATS) axis and balance. Hormones are a beautiful symphony controlled by the pituitary gland within the brain. Your hormone-producing glands act as thermostats that affect your mood, libido, sleep, resilience and more.

The big takeaway to understand when it comes to hormonal imbalances is that when one hormonal gland is overworked, or undersupported (for example, through poor nutrition), it will directly affect the symphony of hormones that is needed for you to feel your best. It is therefore important to look at your hormones holistically with your doctor, rather than in isolation.

If you're concerned that you may be experiencing sustained probelems with your cognition – memory, processing speed, attention, focus; please do discuss this with your GP. There are simple screening tools and tests that can be completed to check your hormonal function.

To take this further into functional medicine, please refer back to the baseline tests I recommended at the end of the previous chapter.

Tech:

HRV – I would like to make a particular mention to Heart Rate Variability as a way to gauge your internal resilience. If you HRV is low (this refers not to the absolute number but the trend for you) it often coincides with feeling less resilient, overwhelmed, anxious, making poor decisions and not being able to give your best. You can measure this with a health wearable such as an Ōura ring or a Polar strap.

Foodforthebrain.org, by Dr Patrick Holford and a global team of cognition experts, is an excellent resource for long-term brain health, with tools for identifying those at risk and simple nutritional and supplement interventions based on evidence.

WHY DO I FEEL SO LOW?

IN THIS CHAPTER, WE'LL FOCUS ON THESE HORMONES:

Serotonin: *makes us feel happy and safe, and regulates our mood*
Oxytocin: *the love chemical that makes us feel cuddly and warm*
Dopamine: *our pleasure hormone, also relates to decision-making, motivation and arousal – low levels can lead to depression, apathy, helplessness*
Noradrenaline/norepinephrine: *works with dopamine to give us energy and alertness – a lack can leave us with low energy and fatigue*

How we feel about ourselves and our life is EVERYTHING. I mean, thank God we talk about mental health more and more these days, and the stigma around depression and anxiety is disappearing as we all begin to understand that struggling with our mood is not something to be ashamed of. So many of us battle with a chemical imbalance in our brain that's down to hormones (of course!) and this imbalance can tip us into a depressed state where life just feels horrifically hard.

In this chapter we're going to explore why and how that happens, but I want to be really clear; this part of the book is NOT just for those of you who have a formal diagnosis of depression. Low mood can impact *all* of us at one time or another – from feeling completely unmotivated and flat to having no energy and just an overwhelming sensation of 'meh'. When we feel like this, life feels like a slog. I know, I've been there: both through regular cycles of crappy moods and also in the horror of depression.

HOW MY MENTAL HEALTH ISSUES DOMINATED MY LIFE

As those of you who've read my first book *It's Not A Diet* will know, I was diagnosed with bipolar disorder after I had my first son, Gray. I was feeling suicidal, completely miserable and was desperate for help. Without looking into my medical history of fertility treatment or doing a single hormonal test on me, my doctor at the time put me on super-strong medication for depression and bipolar, which I then stayed on for the next few years.

I'll never know for certain, but I believe now that the root cause of my mental health problems was my IVF treatment. I'd been pumped full of artificial hormones so that I could get pregnant, and the horrendous after-effects when they dropped off were, I think, behind my terrible postpartum lows. I'd always been outgoing and upbeat before IVF, but once I started messing with my hormones all I could do was panic all day.

Combined with my mental health problems, of course, was my alcohol addiction. I'd already tried a few times to cold turkey it myself, which was a disaster. One time I even went into an alcoholic seizure and ended up in a London hospital. Even then, all I could think was 'Oh look, there's a pub opposite . . .' You are so addicted it's *phenomenal.* It wasn't until I went to a rehab clinic in South Africa that I got sober, which was the first step on my long journey to full recovery.

Although I'd managed to quit alcohol, I spent the next few years feeling low and indifferent to life. I couldn't get excited about anything, because the medication I was on had completely numbed me – I didn't know it then, but I had what's known as *anhedonia.* This is a condition where you are unable to feel pleasure. Thanks to the medication I couldn't cry, but I couldn't really laugh, either. Everything felt flat and kind of pointless. But I was absolutely terrified of going without the medication. I just thought 'anything is better than being in active addiction'. Of course, by then I'd become addicted to sugary processed foods, but I didn't see it like that!

Being on depression medication can be really, really lifesaving for many people (and I'll talk about that in a bit), but for me, it made me totally flatline. I took no joy from life, and I was constantly chasing little dopamine highs with crappy food or shopping for clothes, which I *hated* because I was so overweight. I was really fatigued but couldn't sleep well. Everything became like wading through mud.

It wasn't until years later that I came off the prescription drugs – very slowly, I should say, and under the guidance of a new doctor. I really don't recommend stopping any prescribed medication yourself, without the support and advice of a medical professional. You will need handholding to navigate this properly and minimise the rebound effect.

Getting off the medication was a revelation. I started to feel like myself again for the first time in years. Life went back into full colour and it was another big step for me in the right direction. But even then, I had no idea I could help myself become happier with nutrition and that sugar, veg oils and ultra-processed foods actually triggered my low mood. But as you know, I started to learn more and more about how much our bodies and brains were at the mercy of our hormone levels and what we could do about it.

DISAPPOINTMENT VS ACTUAL DEPRESSION

It's really important to distinguish here between clinical depression and low mood. The lines can be really blurred sometimes between feeling down and serious medical depression, because essentially they're on the same sliding scale of symptoms. I do think the word 'depression' gets bandied about a bit too much, though – people say, 'Oh I'm so *depressed*' whereas in fact, they might just be disappointed! You're allowed to be gutted if you didn't get that job offer (and even have a couple of days feeling rubbish), but don't say you're depressed – you're not.

Clinical depression presents with many different symptoms, from feeling deep grief and sadness, a lowered threshold for crying, anhedonia, to overwhelming guilt about the past or about yourself – delusional negative thinking that you are a really terrible person. You can be utterly exhausted and yet unable to sleep, as you're pumped with anxiety-spiking cortisol, have low libido, and your appetite can dramatically increase or decrease.

A really great analogy about the depressed mind was one I heard on one of my favourite podcasts *The Huberman Lab*. A depressed athlete will say – and completely believe – he's not making any progress in recovering from an injury, even if his physio says he absolutely is. It's that disconnect from reality that is a marker of clinical depression, because it has literally switched off your autonomic nervous system (the system that regulates our bodies' involuntary functions like heart rate, breathing, digestion, sleep, etc.).

A note on the new research about anti-depressants

While I was in the process of writing this book, new research about depression hit the headlines and really made a stir. Scientists from the University of London published a major review that called into question the evidence that depression is caused by a chemical imbalance – and therefore, that taking antidepressants to correct this didn't actually work. However, many other experts – including those from the Royal College of Psychiatrists – disagreed and urged people not to come off their medication without proper supervision.*

Of course, I'm not a doctor, and I can't speak for people's personal experience with antidepressants. Many people consider

* https://www.theguardian.com/society/2022/jul/20/scientists-question-widespread-use-of-antidepressants-after-survey-on-serotonin

that they're lifesaving; but one thing I do believe, is that they're handed out too quickly these days, like they were with me. My friend Becky went to the doctors recently with clear symptoms of perimenopause – including low mood and sleep problems – that she knew *wasn't* clinical depression because the symptoms shifted throughout the month in line with her menstrual cycle. However, the GP still tried to palm her off with a prescription for antidepressants. I hear this sort of story over and over again, so I do believe we have a problem in our society with handing out pills like sweeties. And I believe taking antidepressants made me feel worse because they clogged up my liver, resulting in low dopamine – adding to the triggers for sugar cravings and therefore weight gain.

WHY IT'S ALSO ALL ABOUT OUR CYCLE

I might not struggle with serious depression anymore, but low mood and malaise are still a part of my life. I've learned my mood is definitely influenced by my menstrual cycle (which we'll explore in more depth in the *WTF Is Up With My Hormonal Cycle?* chapter, so definitely take a read of that, too – but it is also relevant here). I definitely get very low around days 2, 3 and 4 of my cycle and, even knowing what I know now, I still believe when I'm in the thick of it that these horrible feelings will last forever! Honestly, I feel as if I've lost 'me', have no motivation or confidence, and won't post on Instagram because I feel completely uncreative and have no idea of what to say. At home I become a bit mute and retreat into myself. Matthew will say, 'Oh, in a couple of days you'll feel OK again,' and I don't believe him, but then snap, like magic, I do. It's bonkers.

Women aged between thirty-five and fifty are those reporting the most mental health issues – from generalised anxiety to depression – so there's a lot of us out there dealing with these problems. I hope knowing you're not alone is helpful in itself. There can be so many factors behind our low moods: from job stresses, to dealing with ageing parents, young kids, money worries, low self-esteem, the list goes on – they all have a knock-on effect on our hormone levels and therefore our mood. But sometimes there ISN'T something major to pin your concerns on and, even if you're not dealing with any of these external pressures, it doesn't mean you still can't struggle with low mood. You don't need my 'permission' to feel a bit shit or need an official clinical depression diagnosis. All of us could and *should* feel better.

Over the years, I've learned so much about how we can boost our mood through understanding which hormones affect how we feel about ourselves, and then what we can do about it. I've been my own guinea pig and found what works for me. I've learned that we can hack our mood hormones positively with diet, supplements, exercise and even some surprising protocols so we feel better all round. And I'm going to show you how. But first let's explore the hormones that we're about to hack.

THE HORMONES THAT AFFECT OUR MOOD

Serotonin

Of course, serotonin is going to be one of the major mood-affecting chemical messengers in our body – after all, it's not called the 'happy hormone' for nothing! If you've read the previous chapters, you'll know that serotonin a) works as both a neurotransmitter (in our brain) and as a hormone (through our bloodstream), b) is made in the pineal gland, and c) is manufactured in our gut (95 per cent of it!).

Healthy serotonin levels are absolutely crucial for regulating our mood. When we have low serotonin, we don't feel joyful or safe in our lives and this feeling can often tip into clinical depression. In fact, most anti-depressant medications are SSRIs and work by increasing levels of serotonin in the brain. SSRIs stands for *selective serotonin reuptake inhibitors* and they prevent our body reabsorbing the serotonin we need to feel well.

The good news about serotonin is that you can naturally supplement your levels with lifestyle and dietary changes. Because the vast majority of serotonin is made in our gut, having a healthy gut microbiome is vital, and there are lots of ways we can help it balance out.

Dopamine

Here's dopamine again, that tricky character that rules both pleasure and pain. We've already seen dopamine play a major role in regulating our sleep, appetite, brain fog and feelings of overwhelm, and here it is again affecting our mood. It's both a hormone and a neurotransmitter as it's behind so many functions and is produced in the brain.

Low dopamine levels are linked with depression, and especially anhedonia, where you lose pleasure in everything (I had anhedonia when I was on antidepressants, as I mentioned earlier), and life feels completely 'meh'. Plus, it's worth noting again that we all have different baseline dopamine levels which determines how we generally feel. You'll probably know those sorts of people who are naturally very chipper and energetic, right? They're likely to have a higher baseline level of dopamine, compared with some other people who are a bit less get-up-and-go-and-let's-change-the-world. I can see differing dopamine levels with my boys. Number 1 has quite low dopamine and he's happy to stay at home and do not much whereas Number 3 is always enthusiastic and wants to go out and play.

As dopamine is a key hormone for motivation and our reward system, when it's out of whack, it can trigger low mood or depressive symptoms like no motivation, difficulty focusing, feeling hopeless and lack of interest in things you used to enjoy.* We need to take really good care of our dopamine levels, as just trying relentlessly to boost them over and over again will deplete our baseline level and leave us feeling worse (for more detail on this, read the *Why Does It Feel Like I'm Losing My Mind?* chapter).

Noradrenaline (norepinephrine)

Noradrenaline, which we've met before, is essentially adrenaline in the brain, and is important for our alertness, energy levels and brain function. What's more, our bodies use dopamine to create noradrenaline, so guess what, if our dopamine is low, so is noradrenaline.

One experiment† (on mice, sorry, don't shoot me) showed that depleting their levels of noradrenaline meant depressive symptoms came back. So if you feel like you can't get out of bed in the mornings and are completely knackered, even if you've slept OK, you may be low on noradrenaline.

* https://www.medicalnewstoday.com/articles/326090#relationship

† https://www.ncbi.nlm.nih.gov/pmc/articles/PMC3131098/

Oxytocin

Oxytocin is a fantastic chemical, often called the love hormone because of the warm, cuddly and fuzzy feelings it gives you. It's produced by the hypothalamus in your brain, and it controls everything from sexual arousal to trust, romantic feelings and bonding. You get a flood of oxytocin when you cuddle your favourite people (or even your pets – I definitely get an oxytocin boost when I see my dogs after being away from them!). One very interesting (cough) fact is oxytocin also gets released when you orgasm, and apparently women produce more of it than men.

Oxytocin is famous too, for being the hormone that bonds mothers with their babies – it's produced to get your womb ready for labour, you get a rush of it after giving birth, and if you breastfeed, too. I definitely felt it with my youngest son, Jude – he was the only one I was able to breastfeed because by then I wasn't on the super-strong medication that meant I couldn't feed the others. I didn't realise it at the time, but when I stopped breastfeeding (after my boobs got engorged – OW), I felt a dramatic drop in my mood. I didn't have that warm glow anymore, I was out of the fuzzy, lovely baby bubble. That was definitely the oxytocin leaving my system!

The important thing to know about oxytocin is that you need somebody else to trigger it – it's a hormone of social bonds. When you're with someone you really care about – whether a friend, family member, child, pet or partner – that sparks oxytocin into action. When you feel crap, you want to shrink back and retreat into yourself, and avoid social connections. I get it – that's me a few days a month, too! – but it's completely counter-productive. Not connecting with others makes your oxytocin levels drop even more, and so it becomes a vicious cycle of low mood.

A side note on cortisol

I think it's really important to mention cortisol again here. It's our stress hormone, produced in the adrenal gland. If our cortisol is out of whack – spiking at the wrong times, or constantly being released thanks to our stressful lives – it can really disrupt our mood. Remember, cortisol is a great motivator, but it also drives fear, and that can have a big effect on depression and low mood.

We all have our foot down on the gas pedal in life with things that stress us out – kids, work, ex-husbands (you know the drill) – so what can you do? Plenty, I promise. I'll explore more about how to manage our cortisol levels in order to manage our stress and anger response in the next chapter, but it's also a good one to take note of here, because it does affect our mood.

A side note on oestrogen

It's a proven fact that women are at higher risk than men of developing mood disorders, and that's all down to our fluctuating oestrogen levels throughout the month.* There are oestrogen receptors on cells throughout our body, which means changing levels of this hormone can lead to mood swings, anxiety, tearfulness and loss of confidence. We'll explore properly how it shifts throughout our menstrual cycle (and during perimenopause and menopause), in the chapter *WTF Is Up With My Hormonal Cycle?* as well as *Where Is This Rage Coming From?*

* https://www.ncbi.nlm.nih.gov/pmc/articles/PMC3753111/

HOW THEY ALL WORK TOGETHER AND MESS WITH OUR MOOD

Even if all these hormones are only a little out of whack, they can have profound effects on your mood. Because, of course, they all work together in a super-complicated dance, affecting each other. On the positive note, a good example is how oxytocin works: when it's released, and you get that lovely fuzzy feeling, it also naturally blunts cortisol levels.

For instance: you're on a night out with your favourite girlfriends, and you're having a brilliant time. You might have been dealing with loads of work stress that day, but suddenly it doesn't matter because that oxytocin has done its job and dampened down the stress hormone. Fabulous, right?

On the negative front, when you're feeling low it's also because loads of hormones are bringing each other down. In clinical depression, three major chemical systems are altered. You'll have less:

1. noradrenaline – affecting your energy
2. serotonin – making you feel guilt, grief and shame, and
3. dopamine – zapping your drive.

What's more, people with depression tend to have elevated levels of cortisol in their bloodstream, making them feel anxious and stressed.

Even if you're not clinically depressed, you can still be depleted in certain hormones like dopamine. If this is the case, you'll have less motivation, and feel 'meh', which means you won't produce enough noradrenaline, for energy. Because of this, you'll be less likely to want to go out and socialise, which means you'll lack oxytocin. It becomes a vicous cycle and you end up feeling worse!

But before you automatically jump on the antidepressant train to tackle your mood, let's take a look at what we can do to help ourselves. Sometimes low mood and out-of-whack hormones can be brought on by things like losing a job, the pandemic, or even lack of sleep, which is why I'd also recommend reading the *Why Can't I Sleep?* chapter, too, if this is you. But you might just be feeling bored and disillusioned with life! Don't worry, whatever it is, we can tackle the root causes and hack our hormones to improve our lives.

SO, WHAT CAUSES LOW MOOD?

Number one – it's inflammation!

Here's a fact that should make us all sit up and go *bloody hell*; chronic inflammation is the root cause of most diseases including depression.* There's a growing store of scientific evidence to prove this too, not least an amazing book that was written a few years ago called *The Inflamed Mind* by a neuroscientist called Edward Bullmore, which brilliantly explores the relationship between our immune system and our mental health.

So, what's going on? It's actually really important to remember that inflammation in the right context is a GOOD thing. Short-term inflammation is lifesaving and necessary especially when it's part of healing. You get inflamed when you get a cold, or any

* https:/bmcmedicine.biomedcentral.com/articles/10.1186/1741-7015-11-200

injury or infection, as the body's immune system rushes into action to try and heal itself. It's an essential way that our body defends itself against foreign invaders like viruses and bacteria, too. So we shouldn't demonise inflammation entirely – without it, we'd be dead!

When inflammation becomes a problem is when it becomes chronic, i.e. long term, and in excess. There are lots of factors that can cause excess inflammation, from stress, to lack of exercise, obesity, poor sleep and OBVIOUSLY what we eat.* I chatted about the foods that can cause inflammation in the *Why Can't I Stop Eating?* chapter as they have a disastrous effect on our appetite-regulating hormones. But it won't surprise you to know that, naturally, inflammation has a mega effect on our mood and risk of depression, too. This is mainly down to the presence of cytokines.

Hang on, WTF are cytokines?

Cytokines are peptides (small proteins) that are produced by our cells and regulate various inflammatory responses in our immune system. So far, so good, especially if we're dealing with an illness or injury. But if we're chronically inflamed, we end up with too many cytokines flying around our system – and they inhibit the release of our mood-regulating hormones: serotonin, dopamine and noradrenaline. They can even limit the amount of serotonin created in our gut. ARGH.

You can see where I'm going with this, can't you? In the right context, cytokines are vital, but when we're ramping up unnecessary inflammation in our brain and body, these

* https://bmcmedicine.biomedcentral.com/articles/10.1186/1741-7015-11-200

cytokines are going to detrimentally impact our feel-good hormones.

(It's worth knowing too, the link between cytokines and Covid. A viral infection like Covid produces an aggressive inflammatory response in our body where the cytokines swing into action to fight it off. People suffering with long Covid symptoms are now thought to be dealing with too many remaining cytokines, which keeps the immune system switched on even when it's not needed.)*

Root cause of inflammation #1 – what you eat

If you haven't read it already, I'd recommend going back to the chapter on *Why Can't I Stop Eating?* for a deep dive into inflammation-spiking foods, and what the worst culprits are for this. In a nutshell, they are:

- vegetable oils (sunflower, rapeseed, etc.)
- emulsifiers
- refined sugar (fruit and honey are OK as they contain enzymes that metabolise their sugars differently)

The big problem is pretty much all of the above are in all the hyper-processed foods we have as part of our regular crappy Western diet. They cause a cycle of inflammation and addictive eating that is really hard to break when you're caught in those hormone-fuelled cravings. I know because I've been there – and I still am sometimes! I feel so low and flatline after I've eaten a Chinese takeaway that's been cooked in veg oil. I see these sorts

* https://www.forbes.com/sites/williamhaseltine/2022/01/25/new-clues-to-long-covid-prolonged-inflammatory-response/

of effects with my followers on Instagram, too. Bev commented, 'Food is my comfort blanket. When I'm down, I turn to sugar to get me through it. It's such a negative relationship and one I'm seriously trying to break.'

Trying to stick to a low-fat, low-calorie diet won't help your mood either, as your body needs good fats to produce your happy hormones. Eating low-fat is super detrimental to our mental health and I'll outline in a bit the mood-boosting foods and supplements we SHOULD be stacking our diets with instead.

Root cause of inflammation #2 – life stress

There's a clear causal link between stress and depression or low mood, and that's down to cortisol. Too much cortisol in our system messes with serotonin production and leaves us constantly in that 'fight-or-flight' fear mode that is so conducive to feeling down.

We're all wired to deal with a certain amount of stress in our lives, but constantly putting our bodies under huge stress has been shown to increase our depression risk. For the average person, around four to five bouts of serious life stress – such as bereavement, losing a job, a relationship breakdown, etc. – will mean you're more likely to suffer with depression because you're impacting your neuromodulators. Your poor body will be pumping out cortisol non-stop as it's perceiving danger and threats constantly.

Of course, we can't live a completely stress-free life, meditating in the forest and omm-omming. We can't quit our jobs *just* because they're stressful – after all, we all live in the real world, we need to earn money and we need to deal with less-than-perfect situations. But you can put boundaries and helpful protocols in place and remove some important stressors from your life before you start to burn out.

Root cause of inflammation #3 – poor sleep

Like I said right at the start of this book, good sleep is the bedrock of our overall health. If you're not getting enough, or enough good *quality* sleep, you're going to feel like shit and it'll have a knock-on effect on pretty much everything – and that includes low mood and depression.

Research from Harvard university shows that cytokines and other signs of inflammation have been found in people who sleep poorly.* So not getting enough sleep can definitely have a negative impact on your mental health. I'll give you a couple of ideas to boost good sleep later on, but you should also read the *Why Can't I Sleep?* chapter for a really deep dive.

HOW OUR GENES AND NATURAL LIGHT COME INTO PLAY

It's long been known that lack of sunlight, especially during the winter months, can make us feel low and lethargic. The reason for this is it knocks our cells' natural circadian rhythm out of whack, which you'll remember if you read the chapter about sleep (don't worry, I'm not going to test you!). The lower light levels of winter disrupt our body clock and levels of serotonin production so we end up feeling more tired and low than usual.

If your mood feels significantly worse during the winter

* https://www.health.harvard.edu/sleep/how-sleep-deprivation-can-cause-inflammation

months, you may be dealing with SAD (seasonal affective disorder),* but lack of light can affect your mood at ANY time of year – especially if you're not getting enough natural light from outside. Light plus cholesterol in the skin enables serotonin to cross the blood–brain barrier from the gut, making you FEEL this happy hormone mentally.

Unfortunately, too, some of us are just more prone to low mood than others. Scientists have discovered a particular gene called 5-HTTLPR (I know, trips off the tongue, doesn't it?) that is a marker for lower serotonin levels in the brain. What this means is that if you have this gene you're chemically wired to be more prone to depression, and may only need one or two bouts of stress to tip you into a clinical low. If you have this marker, antidepressants can really help rebalance your serotonin.

HOW TO HACK YOUR MOOD POSITIVELY

I get how impossible it can sometimes feel to help yourself when you're battling low mood. You're caught in a cycle of doom, and even if you're not clinically depressed (and again, please go and speak to your GP/specialist if you think this might be you) well-meaning people just telling you, 'Oh, go out and have a run' is no use. Sometimes you can feel like, 'I can't get out of frigging *bed*, love, I would rather die than go for a run.'

But it's all about taking small steps, one at a time. I'm going to suggest a few simple, actionable protocols that you can apply to different parts of your life: firstly, what you eat and supplement

* https://www.nhs.uk/mental-health/conditions/seasonal-affective-disorder-sad/overview/

with, secondly what you can do at home, and thirdly, things you can do out of the home. Like with all biohacking tips, try a few, track a few and see which works best for you – even a little positive change is a good one!

Repair and restore with NUTRITION

First things first – you've got to cut out as many inflammatory foods as humanly possible. It might feel impossible to start with, but like I said before, you aren't telling yourself 'I'm never going to eat Pringles again' – it's just for now. You might think you're removing all your 'treats' like biscuits, crisps and chocolate, but I promise you, these aren't treats – they're making you feel worse in the long term. See my previous book, *It's Not a Diet*, for a reset plan to help you ease into the Paleo lifestyle that has really worked for me.

Spice up your life

There are loads of herbs and spices that you can add to your food and drinks to reduce inflammation. These include:

- ginger
- turmeric
- cinnamon
- parsley
- garlic
- cayenne pepper
- black pepper

One amazing thing I learned recently is that you can even

increase the bioavailability of turmeric by adding black pepper! It works like an adjuvant and actually increases the turmeric's efficacy, which is an amazing easy win.

I double down on these herbs and spices when I feel low (usually when I'm due on). Sometimes I even whizz up my very own Mood Shot: a chunk of chopped ginger with lemon juice and a small amount of water (so it's the size of a shot of alcohol!), which I blend together then neck back. It really dampens down the cravings for carbs which comes from low dopamine.

Eat fatty acids to balance your brain

We need good fat because it provides the essential 'building blocks' for our brains, cells, neurons and hormones – it's completely different from the fat around our bellies, in case you haven't got that message already in this book! You know I like a good rant about why fat is good for you.

The science 100 per cent supports eating good fats. Foods high in omega-3 fatty acids have been found in research to reduce the risk of developing depressed mood,* and certain cultures have already factored this into their traditional diet. I referenced this in my last book, but it's worth mentioning again; Scandinavians, who deal with very low amounts of sunlight during the winter months, have a very fish-dominant diet, which boosts their omega-3 intake hugely, and helps mitigate seasonal depression. This isn't something they've just suddenly decided to do because of the science, it's how they've naturally evolved over thousands of years to deal with the dark winter

* https://www.psychologytoday.com/us/blog/integrative-mental-health-care/201812/omega-3s-depressed-mood

months and naturally boost their happy hormones.

Foods that are super-high sources of omega-3 fatty acids are:

- mackerel
- salmon
- herrings
- oysters
- sardines
- caviar
- flax and chia seeds
- walnuts

Take a read of the *Why Can't I Stop Eating?* chapter, too, for a good selection of other happy hormone-supporting foods.

Boost your gut bacteria

A healthy gut microbiome means healthy serotonin levels, which means better mood. Fact! There are more and more studies out there that support how powerful the brain–gut axis is in all aspects of our wellbeing, and I think we're just at the beginning of understanding just how much the connection between our gut and our brain impacts us. One fascinating study on humans showed that just a month of taking probiotic bacteria supplements led to a decrease in anxiety and depression.

This is why I stack my diet with as many fermented foods as possible, and you probably know by now that my drug of choice is kombucha – I can't get enough of it! Other fermented foods are kimchi, kefir, sauerkraut and certain types of live, unsweetened yogurt, so have a try and see which ones you like the best.

My top five mood-boosting supplements

As I've said before, supplements aren't a one-size fits all solution. We can all react differently to the same thing – some of us may find a particular supplement makes no difference, others that it's transformational – so trial and error is your best approach. Try one at a time too, so you can tell if it's having a positive effect or not.

EPA

This really is my number one supplement for low mood. EPA stands for eicosapentaenoic acid (yep, don't even bother), that's found in omega-3 fatty acids. Our bodies need EPA to function optimally, but it's also been found to be hugely beneficial in managing low mood.

It not only reduces inflammation but crosses the blood–brain barrier to get right in there. Studies have shown that an increase in EPA fatty acid can have the same effect as antidepressants. A recommended dose is around 2,000–4,000mg per day.

Creatine

This is a chemical naturally found in the body and is already a popular sports supplement for increasing muscle mass. However, don't be scared – it's great for depressive mood too, and you're not going to bulk up like a maniac just as long as you're not stacking it with steroids and pumping weights in the gym (and I *really* wouldn't advise that)!

Creatine is great for boosting drive and motivation, and in an early trial has been found to have anti-depressant properties.*

* https://www.ncbi.nlm.nih.gov/pmc/articles/PMC6769464/

It works by helping the phosphocreatine in the brain, which makes ATP – energy for your cells. A daily dose can be between 1 and 5g.

Ginseng

You might have heard of ginseng before but been a bit confused, because there are SO many different types, like Korean red ginseng, American ginseng, Japanese ginseng – the list goes on. (Ashwagandha is sometimes called 'Indian ginseng' but it's actually a different plant.) Basically, all ginseng is a type of panax root, and this has been used for centuries throughout the world as a powerful nutritional supplement.

Many studies have shown that ginseng has an anti-stress and anti-inflammatory effect and helps relieve fatigue. A typical daily dose is 200–400mg.

Rhodiola rosea

This is another herbal adaptogen used worldwide to tackle low mood. It has a proper effect on stress-related fatigue and depression as it positively influences levels of serotonin and dopamine in the brain. You can take 200mg once or twice a day.

Mucuna pruriens

This tropical legume has been known for thousands of years to have a 'magical' effect on brain health. We now know that this is because it's a good source of L-Dopa, a precursor to dopamine. A daily dose is between 100 and 150mg.

FOUR WAYS TO RESET YOUR MOOD FROM HOME

Bath for balance

For years, Epsom salt baths have been promoted as a great solution for anxiety, aches and pains and now we know they have a beneficial effect on mood, too. Epsom salt isn't anything like the salt you eat – it's actually magnesium sulphate, and magnesium is one of those super-minerals our bodies need, but often don't get enough of. Magnesium is essential for regulating our cortisol (stress hormone) and increases our calming GABA neurotransmitter levels.

If I'm feeling less-than, I'll stack my magnesium bath with a sauna, followed by a cold shower. I find this really helps boost my mood when I can't find the motivation to get out for a run. It feels manageable when the idea of exercise is too much.

Light therapy

Getting out in the morning to catch some natural sunlight for a few minutes is something you will know by now that I bang on about regularly! Exposure to sunlight is amazing for so many issues, not least low mood, because our bodies convert light to vitamin (hormone!) D, and stimulate the mitochondria in our cells, giving us energy.

However, there's very little we can do about the weather in our lovely country, and when it's overcast, dark and cloudy, we're not going to get as much powerful natural light as we need. That's where red light therapy can help – these devices actually mimic the sun's infrared and near-infrared rays (in huge contrast to our phone screens, which emit sleep-disrupting blue light). You can now buy red light boxes to use at home for a few minutes a day, and not only can they help with seasonal low mood, they're also great for healing and skin issues and good for shift workers. Even just one session of light therapy has been shown to help alleviate depressive symptoms.*

Listen to music

Music is such a powerful tool for boosting your mood and reducing anxiety. Upbeat music has been found to trigger dopamine release, which explains why I love listening to pumping 90s house music when I go on a run!

It's impossible to recommend a specific sort of music to lift low mood because it's just so personal. What's a beautiful song for one person, will be a song someone else can't stand, or the song that another person was dumped to! But uplifting music has been proven to have an amazing effect on hormone production, so it's well worth putting together a playlist of tracks that you really love to listen to when you need a feel-good boost.

Breathwork

I've already mentioned deep breathing as a super-fast super-efficient hormone hack, and it works just as well for boosting your mood, too. Deep breathing activates our parasympathetic nervous system, which triggers relaxing serotonin. It's so accessible, but I'd first recommend following a guided breathwork session online or on an app if you're completely new to it.

* https://www.ncbi.nlm.nih.gov/pmc/articles/PMC5336550/

The other day I was having a hell of a time, so I took myself off to do three rounds of Wim Hof breathing, using his app. When I came downstairs afterwards I felt like a completely different human being. It changed my chemistry within fifteen minutes, I can literally feel the serotonin buzz around my body when I do it.

FIVE WAYS TO SHIFT YOUR MOOD BY LOOKING OUTWARDS

Get out into nature . . .

The power of the natural world is now known to be hugely beneficial in alleviating low mood. GPs in some parts of the UK are now told to offer 'green social prescriptions' to patients with depressive symptoms, rather than handing out antidepressants as a first step. This is about encouraging people to take part in nature-based activities like walking, exercise, gardening, and other things that take place outdoors, and that trigger relaxing, calming hormones like serotonin.

It doesn't mean you have to be surrounded by beautiful open fields to benefit from the natural world – I know most of us don't have that on tap. Of course, it's very nice to be surrounded by sheep and trees if you are (and I'm definitely learning to love my more rural local area up north), but even just pounding the pavements in an urban area can give you the same benefits if there are some trees and plants around.

. . . and use the power of human nature

Feeling better isn't just down to hugging a frigging tree. As we know, oxytocin, the lovely cuddly hormone, is activated by interaction with other people. This is a complete catch-22 if you're feeling really down, because your natural inclination in that situation is to completely retreat from the world. You don't want to reply to texts, to go out, or to engage with your friends and family. I know, because I've been there! But I also know now, that from a hormonal level, this just makes it worse.

We need those social connections to feel good – which is something all of us experienced first-hand during the pandemic, when we couldn't be around each other. You don't have to force yourself to be hyper-social if you can't bear the thought, you just need to have small interactions with others to lift you. I'll give you an example: recently I was out on a run – and *seriously* flagging – when I went past a church. A young guy was standing outside and he was clearly about to get married as he was all dressed up. I gave him the thumbs up sign as I ran past, and his little face completely lit up!

That tiny moment of genuine connection gave me literal goosebumps and a warm glow inside that motivated me to carry on with my run. Human nature is so lovely and endearing, and when you engage with others, it does so much good for your happy hormones. So, whatever you can do to boost your social connections, do it.

Exercise as much as you can

Movement and physical exercise are SO good for shifting low mood – they boost endorphins, which are another one of our happy hormones – as well as encouraging production of tryptophan, which our bodies need for serotonin. Between 60 and 180 minutes of cardio per week is enough to increase the oxygen levels in our brain, and activate brain function which alleviates feelings of low mood. Studies also show that those who exercise more have a reduced risk of depression.*

HOWEVER. I do understand that exercising is easier said than done. If you're dealing with low levels of noradrenaline and dopamine, then exercise might not be available to you yet, and let me reassure you that this is okay. If this is the case, try any of the other protocols I've mentioned here and build up to starting some exercise. It doesn't have to be super challenging, either. Even uphill walking with some music on (I love 70s and 80s disco tracks for a boost of nostalgia!) will make you feel better.

Put in stress-removing boundaries

Cortisol imbalance plays a big role in low mood, as we've seen. If our lives are too stressful, our bodies produce too much cortisol when it's not wanted, and that in turn disrupts production of oxytocin and serotonin. Cortisol is produced by stressful events, and we can't take out all of these, as I mentioned, but we can put in BOUNDARIES to mitigate how stress affects us.

Take a good look at what is causing you to feel low and install some sanity-saving boundaries – are you stressed out by late-night emails from your boss? Is it a particular WhatsApp group that is driving you crazy? Is it having to rush home and sort dinner every night of the week? Do whatever you can to remove the biggest stressors from your life and dampen down cortisol

* https://www.ncbi.nlm.nih.gov/pmc/articles/PMC7415205/

– whether that's archiving WhatsApp chats, putting blockers on your phone, or delegating meal prep to someone else in your house a few times a week. Whatever it is you can do, do it.

Finally, give yourself meaningful goals

Now I don't want to sound all woo-woo here, and like one of those annoying 'reach for the stars!' Instagram posts. But it can genuinely help your mood to give yourself some achieveable goals to help motivate you with your new positive habits. One of the biggest problems with low mood is it often goes hand-in-hand with low self-esteem: you feel shit, so you have no energy to change anything, so you feel shitter and like a failure, the cycle goes on.

You can break that cycle by trying out some of the protocols and tips I've mentioned in this chapter, but make them part of a goal that is personal to *you*. Your goal could be anything, and it doesn't matter how small it sounds. Making massive, insanely ambitious promises to yourself like 'I'm going to be super-positive every single day' are pointless, because you're setting yourself up to fail (NOBODY feels like this every single day!).

When I was overweight and had just started out running a few years ago, I remember my meaningful goal was thinking I would love to have a more traditional type of office-based job. I was living in London at the time, and I remember looking at the skyscrapers in Canary Wharf thinking to myself, 'One day I'm going to work out of an office there.' It might sound silly now, but it was enough to get me jogging round that corner. Eventually, I ended up running the London marathon, but I didn't end up working in Canary Wharf – my career's taken a very firm turn into biohacking, as you know! But it doesn't matter – that goal was what I needed at the time to inspire me and pull me (gradually!) out of feeling so bad about myself.

So, take your time and think about a goal that feels meaningful to you right now. Write it down here and use it as your own personal motivational driving force.

DR E – HOW TO TALK TO YOUR DOCTOR ABOUT LOW MOOD AND DEPRESSION

While everyone experiences emotional ups and downs, it is wise to take action if you are suffering from prolonged low mood, anxiety and depression. In my clinical experience low mood and depression can be a rapidly progressing spiral, so early detection, awareness and intervention is essential. As Davinia advises, seeking human connection, taking time out with your loved ones can help you release the hormones that will bring you back to a sense of safety.

However, if, despite seeking out connection with others, and trying Davinia's interventions, you are still feeling low, lacking motivation; experiencing anxiety, overthinking, or lack of sleep; and especially if you feel helpless and hopeless about the future, then please seek professional help as soon as possible.

It can be helpful to remember that depression, while it may feel like it will never lift, is a temporary state. There is light at the end of the tunnel, even when it doesn't feel like it.

Think about your coping mechanisms at times of stress and strain, and whether they are truly helping. More often than not the unhealthy habits we adopt to cope, such as drinking more or eating junk food, only serve to deepen the depression/anxiety.

Some of the environmental factors affecting our biological base for emotional wellbeing:

- Always being on. WhatsApp, email and message notifications are effectively received by the body as an alarm. With each alarm we have a biological response of cortisol release and sympathetic nervous system activation.

- Social media affecting our expectations of what's normal in life: how we look, things we should have, and vast comparison challenges.
- The unsettled post-pandemic world – financially, loss, stress and fear.
- Fear-based news stories – we are only shown stories of devastation, loss, death and war. Be mindful what media you consume, and don't feel guilty for switching off when you can't cope.

Please seek your local health service's support if you are suffering with any of the following symptoms:

- Lack of motivation
- Feeling worthless or have inappropriate guilt
- Poor concentration or indecision
- Excessive or lack of sleep
- Significant weight loss or gain or change in appetite
- Fatigue
- Thoughts of death or suicidal thoughts
- Panic attacks with or without physical symptoms such as chest pain, breathlessness, heart racing, sweatiness.

Medication can help during the lowest periods, giving you support to help you have motivation to work on yourself.

Low mood and anxiety has biological causes that can be addressed, as Davinia has mentioned above; try these changes in a simple manageable way, and even if it's only one small change a week, it's a positive step towards regaining control of your emoptional wellbeing.

You may wish to work with your GP to take the following tests, which can help address any other health issues that could be causing your low mood and/or depression.

Blood work:

- Hormones: TSH, T4, thyroid antibodies, DHEA, cortisol, oestrogen, progesterone, FSH
- Full blood count: anaemia/B12/folate deficiency, white cell distribution
- Liver function test
- B12/folate/homocysteine – methylation
- Vitamin D – it's a steroid hormone
- Metabolism – blood sugar control and its impact (glucose, insulin, HbA1c)

Integrative medicine:

- Digestive testing: comprehensive stool testing
- Organic acid test: neurotransmitter processing (key nutrients involved)
- Core mineral status

WHERE IS THIS RAGE COMING FROM?

IN THIS CHAPTER, WE'LL FOCUS ON THESE HORMONES:

Cortisol: *the main one that regulates our stress response*
Oestrogen: *the female sex hormone that can cause mood issues when it's dysregulated*
Progesterone: *keeps us calm and dampens irritability*
Testosterone: *regulates our energy and libido, makes effort feel good*
With a note on ***serotonin, GABA*** *and our* ***thyroid***

I am a feisty person, and always have been. I found my voice at a young age: ever since I was a little girl, I have expressed my anger very clearly! Luckily for me, though, my rage doesn't hang around – I explode, and then I'll get over it. In that way, I'm a lot like my dad (and also my number 2 son) – they both go nuclear, and then it's all over. Of course, though, that means we often butt heads. When my dad and I get together, there's usually a huge argument. In fact, it used to be the sign of a successful family get-together, everyone saying the next day, 'Oooh, we had a good row, didn't we?'

We all deal with anger and stress completely differently, and often that's linked to how we were raised and whether we were 'allowed' to lose our temper. My mum was completely different to me: she was really quiet and calm, but despite that, she never suppressed my rage. Looking back, though, she pushed down so much of her own stuff and it was all lying there underneath. She was a bit like that swan – calm and serene on the surface but churning like mad underneath the water!

Expressing our anger safely can be a really helpful tool to have at our disposal. If we're able to get rid of it and move on, it stops anger ruminating as stress, and as we'll see later on, holding on to our anger can have really detrimental effects on our levels of stress hormones. Anger and stress are very linked – I think we can all relate to lashing out when things are getting too much for us! And, as we'll see in this chapter, anger can be a very useful indicator that we need to take steps to deal with the things that are stressing us out.

WHEN STRESS BECOMES TOXIC

Of course, feeling rage from time to time is unavoidable! You simply cannot live in this busy modern world and *not* have some stress going on – and in small amounts, that would be fine.

After all, we're hardwired as human beings to deal with stress. Thousands of years ago we'd release adrenaline and cortisol if we thought we were about to be chased by a predator, as they are hormones that would help us run away faster. But now these hormones are released when we're being attacked verbally with a mardy email, being chased to pay our credit-card bill or being cut up by someone when we're driving. Road rage is a classic one where so many of us blow up immediately, even if normally we're pretty mild-mannered (not me, then).

When I asked my followers on Instagram for their experiences with rage, it was clear that modern life triggers us in multiple ways. Christina says, '[It's] the kids and technology for me. Bloody devices not working properly . . . kids squabbling and fighting with each other . . . talking back to me, being cheeky,' and Laura adds, 'I get angry with the traffic, next door neighbour, loud chewing, partner's snoring, slow walkers, people who don't say please or thank you, the list is endless . . .' Joanne also shares how her wild moods affect her, 'People [trigger me] but mainly the kids not doing as they are told, people who can't drive piss me off . . . I just lose it and I hate it!'

WHY FEAR TRIGGERS A HORMONAL ANGER RESPONSE

But why does this happen? Where's our anger actually coming from? I didn't understand until recently how much of our anger is really about FEAR. When anger surges, it's often because we're scared – one of our primal needs isn't being met and we're catastrophising about what the results of that will be. When I was younger, I didn't understand that my anger was coming from a fear of being late to school, or a fear of losing my friends, or sometimes even a fear of failure.

Going back to the road rage example, we're scared here of the immediate danger that person might be posing to us. If we spot someone using their phone when they're driving, for example, our brain instantly goes on high alert, thinking subconsciously *how is that going to threaten my life?* We'll fast forward to the worst-case scenario (they'll crash into us and kill us), which ramps up our cortisol, the stress hormone. It's cortisol which gives us that panicked, wired feeling and too much of it flooding our system makes us feel horrendous. As Anna on Instagram says: 'It's not one real trigger [for me], it's mostly like a build-up and I feel like cortisol is just seeping from my pores.'

Anger is a normal emotional response when we feel threatened, but twenty-first-century living is making things harder and harder for us, as those cortisol-spiking stressors just never end. The news, kids, families, financial pressures, job worries, social media – these all amp up our stress response, which can tip over into aggression. I'm definitely always on high alert and can easily snap and lose my rag unless I'm in my safe space (home, watching Netflix!). Funnily enough, I've noticed that this is getting worse as I get older. Yes, I'm (hopefully) a bit wiser and less likely to take shit from people, but I definitely can't deal with all the things I used to do in my twenties and still keep my cool.

Back then for me, travel was exciting and fun. I'd get all amped up thinking about what clothes I would wear and what I'd do – but now, just the *thought* of travel and holidays stresses me out! I start fretting *Will I be able to sleep? Will they have Netflix in the hotel so I can chill out and feel safe? Will I look the wrong way down the street when I'm crossing the road and get hit by a car?* For me, it's no longer an enjoyable experience. I pack badly, forget stuff and feel completely overwhelmed and I just want to get back to my dogs and my usual routine. And these days, every time I get on a plane I get a bloody cold sore – a sure sign my body is stressed!

This is because, as we age, our shifting hormonal levels have a huge impact on our stress and anger response, which I'll explore in this chapter, too. Things are made so much worse at different points in our cycle, too. Loads of women on my Insta page shared their frustrations. Karen says, 'I am so angry ten days before my period . . . I know, this is extremely common (as my doctor says) but I'm talking OUT OF CONTROL. I wake up during the night angry or wake up in the morning all angry for no reason. I've cancelled meetings because it's so bad.' By the way, common, or normal, does not mean healthy.

But there are simply times when we can't get to the bottom of WHY our anger bubbles up, too, which leaves us feeling worse. 'Sometimes [I] would have the rage for no reason, usually overwhelm or no control,' says Claire. '[I'd be] like this – patient, patient, patient, boom. No in between . . . I have awful guilt for the times I've lost it with my daughter.'

It's shit, isn't it? Hormone shifts plus modern-life stressors equals anger equals guilt – is there anything we can do about it? Hallelujah, yes there is. We all have our foot down on the gas pedal in life, and there are points where we just cannot take it any longer. If your resilience is on the floor, you're exploding with uncontrollable rage and frustration and you want to gain some understanding and control over your wild moods, then you've come to the right place.

THE HORMONES THAT AFFECT OUR RAGE

Cortisol

Cortisol, which we've met plenty of other times so far, is our body's main motivating hormone, but it's most commonly known as the one behind stress. Yes, it gets us going in the morning – as our main cortisol release is 30 minutes after waking up – but it also drives our fear response, which as we've seen, leads to stress and anger.

Our crazily hectic modern lives are packed with cortisol-spiking activity, because our prehistoric endocrine system hasn't evolved to tell the difference between a real threat to life and an irritating post on social media. These chronic stressors lead to our adrenal glands constantly releasing cortisol, which has a cascade of terrible effects – from inflammation and elevated blood sugar levels, to anxiety, digestion problems, difficulty sleeping and, of course, anger!

High cortisol levels also have a huge knock-on effect on other hormones, especially ... our next hormonal biggie ... oestrogen.

Oestrogen

I can pretty much guarantee that every woman in the country will have at least heard of oestrogen! It's known as the main female sex hormone, which helps develop and maintain our reproductive system. But what most of us don't know (and I certainly didn't for ages) was that oestrogen is a group of hormones that plays a HUGE role in so many more of our body's functions from bone health to regulating good cholesterol, and especially our mood.

All women produce oestrogen throughout their life, but it's

the *type* of oestrogen we make that changes over time. Basically, there are three main types (and very annoyingly they're all really similar words . . .):

- *Oestradiol*: the main one produced by the ovaries, during our fertile years
- *Oestriol*: made during pregnancy by your placenta
- *Oestrone*: produced in the adrenal glands and fatty tissue. After menopause, this is the only oestrogen we naturally continue to make.

When it comes to rage and irritability, our fluctuating oestrogen levels are often to blame, as they're vital for regulating our mood-affecting hormones. But it's not just a matter of dealing with low oestrogen, which is what naturally happens to many of us as we head into perimenopause (and more about that and how the interplay of various hormones work in the next chapter, *WTF Is Up With My Hormonal Cycle?*). Oestrogen *dominance* can be just as responsible for us feeling absolutely FURIOUS and strung out, and I'll explore what these two sides of the oestrogen coin look like in a bit.

Progesterone

This is the one that helps us mellow out! Progesterone is the other major female sex hormone, and it plays a vital role in our menstrual cycle, getting our uterus ready each month for a fertile egg. It's released by our ovaries around the middle of our cycle during the luteal phase, and so progesterone levels peak around then, helping us feel a calm, nurturing softness and a real sense of well-being. It also helps us sleep better and lowers irritability and fatigue.

Unless we become pregnant, however, our progesterone levels then drop away later in our cycle, which can lead to PMS symptoms. That becomes even more pronounced as we head into the perimenopause where our progesterone levels dip even further. If we're low in progesterone, we'll feel anxious and get disrupted sleep, which of course, can lead to rage, because, well, why *wouldn't* it?

Progesterone works in complete synthesis with oestrogen – they are each other's counterpart and work in a see-saw motion. When oestrogen rises, progesterone lowers and vice versa throughout our monthly cycle to keep everything running smoothly. When they're out of whack with each other, problems arise.

Testosterone

We all think of testosterone as the 'male' hormone, but in actual fact, it's actually the most abundant hormone in women too! I know, WTF? It plays a central role in functions like building muscle, burning fat, keeping our metabolism strong and our sex drive healthy (it does this by increasing dopamine). It's a really, really important hormone, and young women produce three to four times more testosterone per day than oestrogen. So testosterone is NOT the enemy, and it's not going to turn you into a man – it makes effort feel good, and increases our ability to do one more rep when we're at the gym.

Testosterone gets a bad reputation because of how we've been conditioned to talk about it as an exclusively male thing. When young lads get all aggressive and get into fights with each other, it's common to put it all down to their 'high testosterone', and not think testosterone has anything to do with us women. But irregular testosterone levels, whether too high or too low, can adversely affect us, too. Conditions like PCOS (polycystic ovarian syndrome) and acne have both been linked to an excess of testosterone.

When it comes to mood, testosterone plays a really vital role, because of how it interacts with too much cortisol (stress) and oestrogen. I'll explain why below.

A note on the thyroid gland

Our thyroid gland is situated in our neck and produces two hormones that regulate our bodies' energy levels, T4 (thyroxine) and T3 (triiodothyronine). It's very common to have thyroid disorders where your body produces too little or too much thyroxine, and these disorders are more common in women than men. You need a healthy thyroid function to make progesterone, but oestrogen dominance, which I'll explore in a bit, can make your thyroid sluggish. I've also discovered very recently that I have a big problem with my thyroid affecting my general health, which I'll explain in the next chapter, *WTF Is Up With My Hormonal Cycle?*

A note on serotonin and GABA

If you've read the other chapters so far, you'll know all about serotonin, our happy, cosy hormone, and GABA, the one that stops us feeling anxious and overly stressed. They're essential for regulating our mood and just being able to get on and enjoy our lives.

Oestrogen – and especially oestradiol, the one we make pre-menopause – actually controls how much serotonin and GABA we produce. Basically, oestradiol fills our tank with serotonin and when that fades away, serotonin levels fade, too, and with it, our ability to control our mood. Without GABA, we can't feel calm. Without serotonin, we don't feel safe and content. Hence – rage. ARGH.

How they all work together to affect our anger

The cortisol steal

Cortisol is known as an alpha hormone, because a bit like an alpha male, it dominates over all the other poor hormones. Our body will prioritise producing cortisol over pretty much every other hormone, which is all very well and good, but the problem arises when we make too much of it. As it's essentially a bit of a selfish bastard, it will STEAL our lovely, calming progesterone to turn it into cortisol, instead!

Sounds bonkers? I'll explain. Progesterone is mainly made in our ovaries, like I said, but a small amount is made in our adrenal glands – the same place we make adrenaline and cortisol. A precursor to progesterone is *pregnenolone*, and that is what our bodies will take and convert to cortisol instead, at the expense of other hormones.

This sounds mad, but it's for really vital, evolutionary reasons; cortisol is the hormone that will save us from a rampaging bear, so our bodies will protect our cortisol pathways at *all costs.* When you're constantly stressed, your body uses cortisol faster than you can produce it, so it looks for other sources to turn into cortisol. Where does it go? Yep, straight to your progesterone precursors, where it 'steals' them for its own gain.

So that's a disaster, right? But not only that, when you have regularly high cortisol levels, it also BLOCKS your progesterone receptors. You may be producing plenty of progesterone but your body will literally not be able to use it. This is why hormonal tests alone can't show us exactly what's going on; we may be *swimming* in progesterone, but if our receptors are switched off thanks to our stress-head friend cortisol, then it's no bloody use to us!

If this cortisol steal is happening, then our stress resilience plummets and we'll feel horrendous: unable to calm ourselves down, anxious and hunted, hunted, hunted! And of course, we're more likely to then explode with anger . . .

The wild oestrogen rollercoaster ride

Oestrogen levels affect the delicate dance of all our lovely mood-calming neurotransmitters, and when they're out of whack – in either direction – we can suffer from intense mood swings. This is really common as we head out of our thirties and towards the perimenopause – for many of us it feels as if we are literally going crazy, with stormy, unpredictable moods that see us go from 0 to 100 in just a few seconds.

Often this can be because we've got depleted oestrogen. Too little of this hormone, as we've seen, means we'll produce lower levels of serotonin and GABA, our feel-good chemicals. With this come disrupted emotions, low mood, difficulty focusing and stress levels that go through the roof.

But although low oestrogen levels are talked about much more these days as a natural evolutionary part of women's life (which is great, don't get me wrong), what we tend to ignore is how high oestrogen levels can be just as detrimental to our mood. Known as oestrogen dominance, if you're dealing with this you'll probably be irritable, raging one minute and weepy the next, have trouble sleeping, put on weight, be knackered and also deal with heavy periods and brain fog. Lovely!

Why excess oestrogen makes us angry

Too much oestrogen circulating around our system (relative to other hormones) is why many of us struggle with feelings of aggression. In order to work properly, our bodies need to detox our excess oestrogen – the 'use it or lose it' idea where we keep only what we need, and flush away the stuff we don't. If everything's going well, our oestrogen pathway would go through these steps (which I've massively simplified from the crazily complicated science!):

1. Oestrogen is detoxed in three chemical stages by our liver, where it binds to certain vitamins, minerals and proteins
2. We excrete (poo and wee) the excess oestrogen via our digestive system (well, of course, where else would we do this?)
3. Hurrah, our oestrogen detox is complete.

However, if you have oestrogen dominance, your oestrogen doesn't get properly broken down and detoxed in this way. You end up with too many useless oestrogen metabolites swimming around your system, which shoves your delicate hormonal balance out of whack and leads to some or all of the symptoms mentioned above.

Too much oestrogen is the baddie – not testosterone!

Something I've discovered recently, which I've found completely mindblowing, is how it's not testosterone that makes us rage-y, it's excess oestrogen. For so long, we've all seen testosterone as the blokey, aggressive hormone (I can't be the only one who associates it with pumped-up guys in the gym) and oestrogen as the 'softer' female hormone.

In fact, it's actually testosterone being turned into oestrogen that makes our moods so erratic – and it's all linked to raised stress levels. Basically, when we're chronically stressed and pinging out cortisol all the frigging time, it raises our levels of a particular enzyme called *aromatase*. Aromatase converts testosterone into oestrogen and then we end up with too much running rampage around our system and messing with our mood. So it's oestrogen that's the killer that makes us feel psychotic and ready for a fight at all costs!

Oestrogen dominance is a real nightmare, because not only does it deplete our healthy testosterone levels, it also suppresses our thyroid function, leaving us utterly knackered. This is because excess oestrogen makes our liver produce high levels of TBG (*thyroid-binding globulin*), which decreases the amount of thyroid hormone available to the body. I know, *cheers!*

So, what causes oestrogen dominance?

Like with any hormonal imbalance, there's never a simple, single reason WHY – wouldn't it be lovely if there was? Instead, there are multiple factors that could be behind why you've got oestrogen dominance, which include:*

- **Poor liver function** – if our liver isn't working well, we can't detox oestrogen properly
- **Taking the contraceptive pill** – it stops our body producing progesterone, which can lead to higher levels of oestrogen
- **Being overweight** – our fat cells make oestrogen, so the more overweight we are, the more oestrogen we produce
- **Toxic chemical oestrogens** – many everyday products

* https://www.thehollandclinic.com/blog/estrogen-dominance

contain dodgy *xenoestrogens* which mimic oestrogen in our body
- **Genetics** – sorry, but sometimes it's just how we're made (though lifestyle changes can help!).

WHY IT'S NOT ABOUT ONE SOLUTION FOR ALL OF US

I'm definitely not down on HRT and will chat about it in the next chapter, *WTF Is Up With My Hormonal Cycle?* But I will say I think the current medical approach to managing oestrogen (and our other female hormones) is wayyyy too simplistic. Too often, women are just told, 'Oh, whack some oestrogen gel on, then you'll feel better!' without even exploring our individual symptoms. It's like Nicola on Instagram says, 'The one size fits all with HRT really is annoying and without testing (as these meno docs suggest isn't necessary) how do we know what we need?'

I agree! Of course, you can have a DUTCH (dry urine) test to check your exact hormonal balance, but I know that isn't available to everyone, as it's quite expensive. And I've definitely made some mistakes trying to balance my oestrogen out. I was in Spain in between lockdowns and feeling really angry and anxious. You can buy HRT over the counter in Spain (I know!), so I got some oestrogen gel. Me being me, I smothered myself in it like it was moisturiser (I know, *I know!)* but it didn't help. I got no uplift in my mood *at all*. I had a similar underwhelming experience with progesterone – taking tablets only made me feel more hollow and scared than I was already. It's completely paradoxical, because progesterone is known for keeping us calm, and for many women, taking the tablets works! But for others, it makes us feel restless and anxious, because it's all about how it interacts with our GABA receptors.

Like with everything hormonal, it's all about trial and error and learning what works for YOU, rather than relying on the 'average' numbers and measurements. So I'd really urge you to look at how you feel, when you feel most angry, write down what your triggers are, what time of the day and month they happen, what's led up to it, etc., all of that. I know it's arduous, but the science isn't there yet to be 100 per cent accurate, so we need to track things ourselves in order to take action!

SO, WHAT CAN WE DO ABOUT IT? IT'S ALL DOWN TO THE DETOX

Whether our oestrogen levels are too low, or too high, the number one thing we can do to help ourselves is help our detox pathways work as well as they can. We can try to balance our levels naturally before we reach for the patches and gels, by hacking our hormones smartly. Oestrogen imbalance causes havoc with our temper and so many of us are struggling with this monster, like Anna on Instagram. 'The rage is crazy and I try over and over again to control it. I hate myself for it,' she posted. I feel your pain – but there is lots we can do to help ourselves feel better and thrive.

HOW TO DETOX OESTROGEN WITH SOME SIMPLE MATHS

There are some really straightforward shifts we can all do to help support our oestrogen detox – and it's a combination of removing certain things, and adding others in. I like to think of it as a really easy maths equation – add this, take away that. First up, let's look at what we should be taking out.

Things to TAKE AWAY to support oestrogen detox

Gluten

Gluten is a known hormone disrupter,* and it can affect oestrogen and progesterone levels. If you're sensitive to gluten, your digestive tract will become inflamed, causing increased cortisol, which leads to it nicking all the other soothing, mood-calming hormones that we saw before. If you're chowing down on gluten and it doesn't agree with you, you'll also stop your body being able to absorb all the nutrition it needs,† as well as slowing down the passage of oestrogen through the gut, which makes your excess oestrogen stay longer in the body.‡

You may not realise you have a gluten sensitivity, but it's

* https://www.glutenfreesociety.org/gluten-sensitivity-hormones-and-vitamins/

† https://medium.com/thrive-global/how-gluten-affects-digestion-and-hormone-balance-for-women-over-40-5a4cecc3ac61

‡ https://www.floliving.com/gluten-and-hormones-is-this-a-problem-for-you/

worthwhile cutting it out for a couple of weeks firstly to see if it makes a difference to your mood. I know the thought of going gluten-free can be painful, especially if you're a bread lover, but there are loads of gluten-free alternatives out there these days (though please read the label to look out for added sugars and/or vegetable and seed oils). Plus, the following grains are all naturally gluten-free:

- Rice
- Quinoa
- Oats
- Buckwheat (yes, I know it sounds like it, but it doesn't really contain wheat)
- Corn
- Teff

Chemical 'zombie' oestrogens

Without realising it, you might have been stacking your life with products containing toxic chemicals called xenoestrogens which mimic oestrogen in our bodies, leading to excess levels. Xenoestrogens are in everything from plastics, to pesticides, cosmetics, sanitary towels and even some till receipts!

There are three main oestrogen mimickers that are found in our everyday products, that get into our bloodstream and disrupt our natural oestrogen.

- ***BPA*** *(Bisphenol A)*: found in cans, plastic bottles, microwave popcorn
- ***Triclosan:*** found in some deodorant, toothpaste, antibac soap and body wash
- ***Phthalate*** (try saying that after a late night): in some

scented candles, perfume, body lotions, air-freshener and plastic wrap.

However, *do not panic*, there are some really easy ways to remove or minimise these baddies from your everyday life:

- Don't buy bottled water! Instead, fill your own stainless-steel bottles with filtered water
- Store leftover food in resealable glass jars (old jam jars are ideal), and ditch the cling film
- Don't heat up plastic containers in the microwave
- Clean pesticides off fresh food (unless you've bought organic) with a tablespoon of apple cider vinegar and a teaspoon of bicarbonate of soda diluted in water
- Use unbleached or organic sanitary products
- Buy make-up and toiletries that are paraben and sulphate-free
- Say no thanks when you're offered the receipt!

Alcohol

Sorry to be a buzzkill, but alcohol is no good for our oestrogen! This is because it's a toxin which our liver then has to metabolise. By drinking alcohol, we're putting extra pressure on our liver, which will affect its ability to detox oestrogen, hence more oestrogen floating about our systems.

It's been shown that even one alcoholic drink a day can ramp up our oestrogen levels and there is also a clear link between breast cancer and alcohol consumption* (if you needed another

* https://remede.com.au/signs-of-oestrogen-dominance-and-how-to-change-it/

reason to ditch it!). If you can't or don't want to cut out alcohol entirely, and I do understand, just try keeping it to the weekends and aim for alcohol-free weeknights.

As I don't drink alcohol at all, I have a nootropic energy drink with L-theanine before a night out to help me with the small talk, and drink kombucha as a wine alternative.

Things to ADD IN to support oestrogen detox

Good fibre

Having a diet rich in fibre is crucial for healthy oestrogen detox. This is because it increases the passage of oestrogen through the bowels. Yes, I'm talking about poo again. If you're constipated, and/or eating mainly crappy processed foods (which you should know by now are a total shitstorm for our hormonal imbalance anyway), then your body simply won't be able to maintain healthy oestrogen levels.

Despite what the marketing people want you to believe, a good source of fibre ISN'T a bowl of All-Bran and it doesn't need to be high-carb stuff like brown bread. Instead, try these to naturally boost your fibre intake (though take it easy, as too much fibre can cause bloating, especially if you're not used to it):

- *Beans:* lentils, pinto, kidney and black beans
- *Fruit:* apples and pears (skin on), kiwi fruit
- *Veg:* carrots, broccoli, cauliflower, watercress and other brassicas
- *Seeds + nuts:* flaxseeds, sunflower seeds and walnuts
- *Grains:* white rice cooked in organic bone broth, quinoa

Turmeric

This potent super-spice counters the negative effects of oestrogen, and it's anti-inflammatory too. You can either mix it with a small bit of water in the morning to have as a turmeric shot or sprinkle a little on whatever you're eating for lunch. Don't forget to combine it with black pepper to make it more bioavailable.

DIM

I will go into proper detail about the incredible compound DIM (diindolylmethane) in the *WTF Is Up With My Hormonal Cycle?* chapter, but it's important to mention it here too. Not only does it balance out your oestrogen levels, but it also stops that annoying enzyme aromatase from working. That's the one that converts testosterone to oestrogen, which as we've seen, can send you absolutely raging. A good dose is 200mg per day.

Phytoestrogens

Unlike the similar-sounding xenoestrogens, which you want to avoid, phytoestrogens are the goodies. They are naturally occurring plant compounds that are really helpful for balancing our oestrogen, and have also been linked to the reduction of menopause symptoms, such as hot flushes.

Foods containing natural phytoestrogens include:

- Fermented soy and tempeh (make sure it's organic, as regular soy uses a lot of pesticides)
- Flax, chia and sesame seeds
- Lentils, chickpeas, black beans
- Carrots and cruciferous vegetables

Get sweaty

A great way to promote your oestrogen detox is by getting your sweat on! However you do it – whether that's exercise, or using a sauna pod – sweating is a fantastic tool for helping ease the toxic burden on your liver and supporting oestrogen detox. Basically, you get rid of all those toxic oestrogen metabolites via the skin, rather than having them hanging around in your system.

You'll have probably already read in previous chapters how important cold water is for all sorts of issues, and I love combining my sweat sessions with a bit of cold, too; ramping my temperature up and down to flush out my system. I have a pop-up sauna most nights followed by a cold shower, which really helps my detox pathways and also gets me feeling really sleepy and ready for bed.

Righto then, over to you. I want YOU to put together your own oestrogen detox sum. Rather than try ALL of these things at once (which could lead to you just giving up – trying to change too many things in one go is bloody hard work), choose one thing to add in and one thing to remove. Write down what you're going to do below:

My oestrogen detox equation

Me + ______________ – ____________ *= more balanced oestrogen*

HOW TO STOP YOUR CORTISOL SPIKING IN SIX STEPS

As we've seen, it's also really important to regulate our cortisol as much as possible, firstly to balance our stress levels, and secondly, to stop it stealing from our other lovely mood-balancing chemicals! There are lots of tips for cortisol balance elsewhere in the book (if you've not already read them, go back and knock yourself out), but these are some highlights I'd like to flag up here:

Take L-theanine

I've already banged on about this supplement loads, but I'm not going to stop here! L-theanine is the perfect supplement to stop your cortisol spiking, because it contains amino acids (naturally found in green teas) which blunt the stress response right down.

I take a lot of L-theanine with my coffee, which for me, is an ideal band-aid situation. It allows me to feel the cool, calm energy from the caffeine without the horrible wired sensation I used to get. Personally, I need caffeine to cope with my life and especially the mania of the school-run mornings. I can't just pack everything in and go and live in a tepee next to a waterfall, so I need a helping hand to manage my stress levels. L-theanine is it.

Remember how I mentioned that I don't enjoy travelling much these days? Well, I recently went to Morocco without any L-theanine (maniac), and I found my stress levels absolutely rocketed. I was convinced the kids were going to get kidnapped, I was completely spiralling into panic. In the end, my dad brought some over for me when he arrived from the UK, which completely sorted me out. For me, it's a non-negotiable supplement!

Heat yourself up

Temperature change is a proven mechanism for balancing our hormones that millions of people are now implementing into their daily routine. When it comes to managing your stress response, heat is your friend. Either spending fifteen to twenty minutes in a sauna at 80–100 degrees,* or a hot bath, followed by a quick douse of cold water, will actually reduce cortisol levels, as well as all the other benefits I mentioned in the section above.

Follow the light

I've already written at length about how important light is for balancing our hormones and especially making sure that we get enough early morning light to promote cortisol. But what's completely fascinating is that our cortisol can be really affected by the time of year we're in.

If you think about lovely long summer days, we often associate that with feeling more relaxed and mellow, right? And that's for hormonal reasons; the early morning light, longer days (and more sunlight!) stimulate our cortisol response better in the mornings, which is when we want it. Your cortisol rises naturally in the morning and then the effect of light reduces overall cortisol levels throughout the day.† So summer days have actually been proven to reduce aggression, and it's not because you're happy having a drink at a bar in Ibiza (although that won't hurt), it's because of the magical properties of natural light on our endocrine system.

* https://journals.sagepub.com/doi/10.1177/15579883211008339

† https://www.ncbi.nlm.nih.gov/pmc/articles/PMC3686562/

In winter, it takes WAAYYYY longer for our natural cortisol awakening response (CAR) to kick off, because it's dark and we tend to hunker down indoors. So, not only will that make it harder for us to wake up and feel energised, it also means we're more likely to spike our cortisol throughout the day, when we don't want it. Not having enough light also affects our vitamin D levels, which is vital for hormone production.

So, if it's dark, cloudy and miserable, this is where a red light box can really help – using it in the morning at those times will really help boost your cortisol when you want it (and dampen it down when you don't).

Take ashwaghanda

Ashwaghanda is a superb adaptogen that has been found in many studies to actually reduce cortisol* and is very useful in situations of low light (which is what most of us have, for most of the year in the northern hemisphere!). What's more, it can help you sleep better. You can take it as a capsule or in powder form, dissolved in a drink (though it's very bitter, so I'd choose the pill). As mentioned before, it can dampen your mood, so pay attention to how you feel if you start using it.

Have a good cry!

Repressing your anger might have been the way you were brought up, but it's a disaster zone for your health – not expressing your angry emotions has been shown to increase blood pressure,

* https://www.healthline.com/nutrition/ashwagandha#1.-May-help-reduce-stress-and-anxiety

insomnia, anxiety and depression, as well as being a higher risk for addiction. If you don't get rid of the excess cortisol and adrenaline flooding your system when you're raging, it's going to hang around and make you ill. It might be socially acceptable to be nice and calm and unruffled all the time, but it doesn't actually work for our hormones!

Instead, have a good cry. Tears actually contain cortisol and help get it out of the body. Whether they're tears of joy, stress or sadness, they all contain cortisol and adrenaline, the release of which helps us to calm down and feel better. So if you want to cry, cry as much as possible, put a weepy film on and let it all out.

Get down to the root cause

When I was in recovery from alcohol addiction, I learned a great acronym that helped me loads. It was FEAR: false evidence appearing real. It's helpful because not only does it help us understand what's behind fear most of the time (something we perceive to be true/a threat, but that isn't really), but also because it's a reminder that most stress and anger is actually fear.

So when it comes to managing your cortisol response, which, at its root, is simply a fear response, I want to challenge you to get to the root cause. Write it down. But rather than just write the first thought that comes into your head – 'Oh, I was angry because that idiot cut me up in traffic this morning' – keep going. Find what your triggers are and keep asking yourself *why* this has such a ferocious effect on your fear response. Step through it until you get to the root cause.

For example, I might kick off in the morning at my kids because I can't find a missing shoe. But I'm not actually angry or scared about a missing shoe – it's what it represents. If I step through my response to get to the root cause, it's usually – shoe is missing – they'll be late for school – they'll miss lessons – they'll fail exams – they won't be able to get a job and support themselves in the future. It's catastrophising, yes, but it's what most of us are doing, albeit subconsciously, when we fly off the handle.

Most of the time when we lose our temper and shout, it's coming from a place of love and affection, that's mistakenly got wrapped up in our fear response. So not only should we dig down to find the root cause of our anger, but we should also explain this to our loved ones! It's taken me years of recovery to work this out – and I think these teachings really should be in school, but that's another issue! What we can do now is be upfront about it – for example, with my kids, saying to them, 'Look, I'm sorry I kicked off. This is how my brain is working – Mum's worried that something bad will happen in about 230 years!' It helps you override the explosion and get back to balance more quickly.

OK, last thing for you to do in this chapter – write down your own cortisol-balancing sum. Try two of the things above and see how it makes you feel. Write down what you're going to do below:

My cortisol balance equation

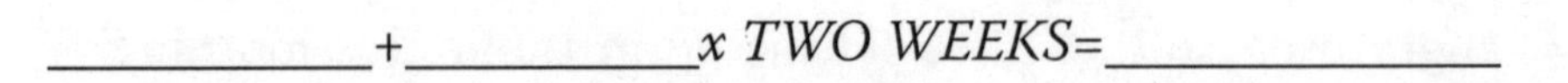

DR E – HOW TO TALK TO YOUR DOCTOR ABOUT RAGE

Stress, and perceived stress, act as a rev to your biological engine – the more you rev the engine, the less efficient it becomes at handling stress over time. Under chronic stress our resilience goes down, which can manifest in anxiety and physical symptoms such as headaches or chest pain.

When you are stressed you may find you make poor decisions which can have an impact on your relationships, career and all aspects of life. It has also been shown that high stress levels are associated with developing chronic diseases earlier in life and with dying younger.

Rather than beat yourself up if you are experiencing bursts of anger or rage, I would urge you to see these outbursts as important signals that you may need some help managing your stress levels. Irritability, rage and snappishness are symptoms rather than personality defects!

I would suggest you should go to see your doctor if you are experiencing any of the symptoms in the list below, all of which may be stress-related. And, of course, be aware of these symptoms in others – it is often hard to notice your stress response yourself, as the increase in stress is often incremental over time and it begins to feel normal. But we all notice if our partner, friend or colleague has become noticeably more short-tempered.

If you are concerned about your rage, there are tools that your doctor can help you access, such as cognitive behavioural therapy and talking therapy. These teach lifelong skills that can help you reframe your relationship with stressors.

Signs your biological resilience to stress is too low:

- Frequently losing your temper
- Violent urges
- Palpitations

- Clammy hands
- Headaches
- Chest pain
- Recurrent thoughts that are troubling
- Misplaced anger
- Restlessness/agitation
- Inner voices commanding or directing you to behave a certain way
- Feeling very suspicious or paranoid
- Being fed back that you're losing your temper too frequently
- If you feel you are at risk of harming anyone and feel a loss of control during any rage
- If you have abused physically, verbally or emotionally, or have felt abused, please seek support.

Services of support via your doctor:

- Behavioural modification support services to help you better understand and prevent downward thought spirals earlier
- Group therapy
- Relationship counselling

Medical tests

Please refer to the medical tests in the *Why Can't I Sleep?* chapter if you want to investigate your anger symptoms further, as many of these tests can help identify and/or eliminate any biological causes.

WTF IS UP WITH MY HORMONAL CYCLE?

IN THIS CHAPTER, WE'LL FOCUS ON THESE HORMONES:

Oestrogen: *the female sex hormone that controls our reproductive system, including our menstrual cycle*
Progesterone: *the other sex hormone which gets us ready for pregnancy (and stabilises irregular menstrual cycles)*
Testosterone: *regulates our energy and libido*
Pregnenolone: *the starter hormone, which can indicate perimenopause when low*
Follicle stimulating hormone: *which causes our ovaries to release oestrogen*
Luteinising hormone: *which stimulates the release of progesterone*

Thyroid hormone: *which regulates our cycle*
With a note on ***cortisol***

I started my period on a rounders pitch when I was almost fourteen. Of course, I'd lied to most of my friends that I'd *already* started by that point, because when I was that age, I was *obsessed* with getting my period, as we all were. I was six months behind my best friend, which, of course, is hardly any difference really, although back then it felt like a lifetime!

It's funny to remember the pressure to hit certain milestones when we were back in our teens – but I think it affects all girls, no matter when you grow up. I was a bit of a late developer, absolutely desperate to have big boobs and big hips and to be as voluptuous as possible. I was at a mixed boarding school on a scholarship, which was great in many ways – as I'd been a lonely only child before then and finally had lots of girls to bond with – but the pressure to be grown up was *ferocious.* Basically, we were all swept along in a race to get a bra, get our periods, lose our virginity, start smoking and get smashed!

As part of that mix, pretty much as soon as we got our periods, we all went on the contraceptive pill. It was seen as *such* a sign of maturity – although, come on, clearly it's the opposite, because it usually means you're not having properly protected sex! But as I said at the beginning of this book, there was no way I wasn't going to go on the pill, despite the fact I had no bloody clue what it was actually doing to my hormone levels, and what potential side effects it could have on me.

For years I didn't give my hormone levels a second thought: not paying attention to when I took the pill, chucking multiple doses down my neck if I missed a few days, carrying on taking it so I wouldn't get my period on holiday, all of that. I was young, didn't suffer from PMT – though I used to pretend I did to get out of PE – and was full of energy, so I didn't think twice about my hormone balance. But then again, I grew up in the 80s and 90s, when everything to do with women's cycles was shrouded in shame and we believed the pill fixed everything!

WHY WE HAVE TO GET RID OF PERIOD SHAME

I still feel really strongly that girls and women are not being taught anywhere *near* enough about what's happening on a biochemical level in their bodies every month. And so much of that has been down to this strong sense of shame that's attached to periods. I mean, it's crazy: our menstrual cycles are as old as time (literally) and are vital for the continued existence of the human species! Without women bleeding every month, there's no humanity, simple as that. But for hundreds of years, periods have been a 'taboo' subject, and so we've been raised to whisper about them and not even given enough information to understand our own bodies.

When I was taught about periods at school, there was no mention of the hormones involved, or what periods were actually *for.* Instead, I just remember a weird cross-section diagram of a tampon going into a uterus, and not understanding how it worked at ALL. Instead, it was always someone's big sister who would tell us the 'real' information, but even then, of course, she had no idea what was going on with our oestrogen and progesterone levels!

It felt impossible to ask questions, and we certainly weren't encouraged to; there was so much shame around periods that had been passed down through generations of women that we were all completely in the dark. I mean, no wonder we didn't worry about taking the pill and messing with our delicate hormonal balance!

WHEN MY CYCLE STARTED TO CHANGE

Once I hit my late thirties, I noticed my PMT symptoms really ramped up. I started experiencing more ferocious rage, my confidence plummeted and my internal monologue became extremely negative. I'd feel no hope, and *what's the point?* about literally everything – myself, my work, my future. I'd be consumed with pessimism and dread, convinced that something awful was going to happen to the kids. This extreme low mood and self-doubt would completely overwhelm me, but then literally a week later I'd feel great – positive and buzzing with energy, so happy with my kids and joyous that Jude had learned to tie his shoelaces brilliantly.

This complete shift in perception from week to week was *phenomenal.* It's the same life, the same bills, the same responsibilities, but a completely different experience of it! The change in mood was shocking to me, but I could see that it was completely cycle-related, *not* clinical depression, because it ebbed and flowed depending on the time of the month.

In many ways, experiencing (and properly noticing) this was a great thing, as it started me down my biohacking route – I became obsessed with finding out what was going on with me hormonally, and whether I was perimenopausal (I was and am! More on the perimenopause later). I researched and read everything I could to understand my hormones and through that knowledge, I got an understanding of my body's cycle.

I cannot overstate how much this knowledge is POWER. Now, even though I do still experience low mood (although it's not as bad as it was), I'm OK with it because I know *this too shall pass.* I understand that when the self-loathing hits me, usually on days 1, 2 and 3 of my cycle, it's because I've dropped in oestrogen and progesterone, *not* because I'm a shit businesswoman, or a rubbish friend and mum! Because I've got that inner knowledge, I can give myself permission to be a bit of a dick to myself, and I know that the mood and procrastination will go in about seventy-two hours so I can bear with it, or better still, hack it.

HOW HACKING OUR HORMONAL CYCLE CAN HELP US

Thankfully, so many more of us are sharing how much our cycle affects us in different ways, and so we know we're not alone in our rage, our cravings, our focus, our mood . . . actually, pretty much every part of our lives. On Instagram Clare says, 'I always know when I'm due on because I turn into a crazed angry person. Everything winds me up then . . .' My friend Jenny even wears a badge when she's due on, so her whole family know why she's acting like a wanker; it's her fluctuating hormones, not them!

In this chapter, I want to give you the knowledge so you can navigate your periods and work around them, whether you're in your fertile twenties or perimenopausal forties! It's vital we understand what is happening with our cycle, and why we might be dealing with ferocious PMT, irregular cycles, insane cravings or depleted energy levels – and what we can do about it. It can often be really confusing to work out what's *really* going on, because so many cycle-related symptoms like brain fog, low mood and raging appetite can cross over into other health problems, and if we're taking synthetic hormones like the pill, it can mask our true symptoms.

But – we're going to do our best to untangle our knotty, complicated hormones and learn how to work *with* our cycle, not against it. Life goes on, and we have to adapt to it; unfortunately, it's not always possible to have a day off when our energy levels are at their lowest (or *not* do a photoshoot when we're on day 2 of our period! Yep, that's me on the cover of this book, great timing, right?).

Instead, when we're armed with hormonal knowledge we'll know when we're going to feel a bit shitter, a bit less energised, and that's fine. We'll learn how to level out our fluctuations, so we don't have the chronic crashes, and hack our hormones positively. Just knowing what's going on with our cycle really gives us the freedom to optimise our lives in the best way we can.

WHY TESTING WON'T ALWAYS GIVE YOU ANSWERS

As I've said throughout this book, hormone testing isn't always the be-all and end-all. You might be suffering from a dictionary's worth of symptoms, but have your blood tested on the NHS and be told that all your results are in the 'normal' range! It's important to remember – again – that 'normal' doesn't mean 'healthy'. Normal just means 'the most common', and we have HUGE ranges between the bottom and the top. Think about it – type 2 diabetes is becoming more and more common, so it might be normal for millions of people (unfortunately), but it certainly isn't healthy.

So, if you've gone to your doctor with horrendous

symptoms of PMT, irregular cycles or wildly changing moods and your blood tests say that you're 'normal', do not be dismayed. The tests can't show everything, least of all your nuanced hormonal levels. If you don't *feel* normal – by which I mean healthy, thriving and energised – and you think your cycle might be behind it, there is lots you can do yourself to help.

Which hormones affect our cycle

Oestrogen

This is one of the two major female sex hormones, and we've already seen in previous chapters how essential oestrogen is to all sorts of functions as there are oestrogen receptors all over our bodies! Oestrogen causes our reproductive system to develop in puberty and it regulates our menstrual cycle, too. It's produced by our ovaries and causes our uterine lining to thicken again each month after our period, reaching a peak at ovulation.

As mentioned earlier, there are three main types of oestrogen that our bodies produce – a quick recap in case you skipped that part:

- *Oestradiol*: the main one produced by the ovaries, during our fertile years
- *Oestriol*: made during pregnancy by your placenta
- *Oestrone*: produced in the adrenal glands and fatty tissue. After menopause, this is the only oestrogen we naturally continue to make.

When our oestrogen levels are unbalanced, this can cause a whole host of horrendous symptoms, as well as wrecking the regularity of our cycle. Oestrogen dominance can be an issue (which I go into in depth in the *Where Is This Rage Coming From?* chapter), and it can also make our periods really heavy and long. Too little oestrogen can also cause horrible effects, like very irregular or light periods, as well as tender breasts.

Progesterone

Progesterone is the other major female sex hormone and works in harmony with oestrogen in setting our menstrual cycle. Like oestrogen, it's also secreted by the ovaries and is produced after ovulation by the empty follicle once our egg has been released.

Its job is to maintain the thick uterine lining ready for pregnancy, but if fertilisation doesn't happen, progesterone levels drop off sharply towards the end of our cycle. However, if pregnancy does occur, progesterone levels stay high, as they're important for maintaining the placenta.

If your progesterone levels are low, you might suffer with water retention and bloating, anxiety and irritability, as well as night sweats, poor sleep, irregular cycles and fertility issues.

Testosterone

If you've read the chapter *Where Is This Rage Coming From?* you'll already know how essential testosterone is for women as well as men. You might also be surprised to learn it has a role to play in our menstrual cycle too (well, I certainly was, but you know, I love this stuff).

Unlike oestrogen and progesterone, testosterone doesn't regulate our menstrual cycle, but instead controls our sex drive. Combined with a rise in oestrogen, we're likely to feel more up for it around ovulation,* because, like it or not, our bodies are hardwired to want to reproduce!

Too much testosterone floating around your system can cause acne and skin problems, increased body hair, and issues with sleep, as well as polycystic ovarian syndrome (PCOS).

Pregnenolone

Pregnenolone is the 'starter hormone' for loads of hormones essential for our cycle, such as oestrogen, progesterone, testosterone, plus cortisol and others. Produced in our adrenal glands and made from cholesterol (because remember, we NEED cholesterol!), it's sometimes known as the *precursor hormone.*

Low pregnenolone levels are a handy flag that we're heading into perimenopause and can be behind symptoms like brain fog and insomnia. Luckily, unlike poor old melatonin (our sleep-supporting hormone), we *can* buy pregnenolone as a supplement in the UK without a prescription.

Follicle Stimulating Hormone (FSH)

This hormone is produced in the pituitary gland (in our brain). It causes our egg to mature in the ovary and also stimulates the ovaries to release oestrogen.

Luteinising hormone (LH)

Also made in the pituitary gland, luteinising hormone causes the mature egg to be released from the ovary, and stimulates the release of progesterone.

* https://moodymonth.com/articles/hormone-101-testosterone

Thyroid

Our thyroid gland is shaped like a butterfly and is at the base of our throat. The thyroid helps control our menstrual cycle, and so if our thyroid levels are out of whack, so's our cycle. Too little thyroid hormone (hypothyroidism) can cause heavy bleeding, while too much (hyperthyroidism) can lead to light, or even absent periods.

A note on cortisol levels

Cortisol is one of those hormones that just keeps on popping up, doesn't it? What's good to know here is that our cortisol levels gradually increase with age, which can put extra pressure on our adrenal glands, causing what's known as *adrenal fatigue.* Symptoms of adrenal fatigue are mood swings, heavy bleeding, headaches and increased PMS – essentially, many of the symptoms when our OTHER hormones are dysregulated. Of course, this means we can be suffering from adrenal fatigue without realising.

SO, WTF IS GOING ON WITH OUR HORMONES DURING OUR MENSTRUAL CYCLE?

Puberty

Everything cycle-related is kick-started when we go through puberty, which can be from any age from eleven or so onwards. Female puberty is defined as the period when we mature physically into a woman, which is triggered by good old oestrogen: we grow body hair, our hips widen and breasts develop. We also start our periods.

Fertile years and the pill

Once we've started our period, we're then in for a good few decades of it, before perimenopause starts to shift things around again. We go through a monthly cycle where our hormones ebb and flow, depending on what phase we're in, and as we all know, this can have huge effects on our physical, mental and emotional wellbeing.

This is where the contraceptive pill starts to feature for many of us – and not only for preventing pregnancy. Like I mentioned before, it had massive social cachet for my peers back in the 90s, and I know loads of women who say the same. Taking the pill can also help regulate periods, lessen menstrual pain and reduce acne; in fact, one in three teenagers have been advised to take the pill for reasons *other* than birth control, so there is a lot of pressure out there to get on it. But too many young women don't fully understand what exactly they're doing to their delicate hormonal balance when they start on the pill – and we already know that I *certainly* didn't.

What the pill does to our hormones

The combined pill contains artificial oestrogen and progesterone to mimic pregnancy, as well as a host of other hormones. In fact, there are nine different hormones involved in the pill, so essentially, you're already on HRT if you take it! Not only will it mask any genuine hormonal imbalances you could have, it can cause all sorts of unexpected health issues that I believe we're simply not educated enough about.

Research at the University of Copenhagen found that teenage girls on the pill were 80 per cent more likely to be diagnosed with clinical depression and there's an increased risk of cervical and breast cancer for women who've been on it for a decade or longer. It interferes with the body's stress response, and can even influence our choice of partner, according to Sarah E. Hill's *How the Pill Changes Everything.* Scary stuff. The pill also affects the body's absorption of essential vitamins and minerals, can impact our fertility and increase risk of blood clots, too.

It is crazy how little of this we actually know before we start chucking hormonal birth control down our necks and feeling all super-mature. I'm not here to say, 'Don't use the pill,' I just believe we need to be far more informed about what we're doing to our hormonal balance and how the pill can knock it out of whack. If you're reading this and you're still young, you still have a window to find out what your optimal hormonal balance is. I'd recommend doing a DUTCH test to see what YOUR optimal levels are to refer back to over the next couple of decades. I wish I had this to refer to from my twenties. Measuring your hormonal baseline is a great reference point for the future, and a perfect gift for your daughter (help the next generation avoid the guessing game).

If you are interested in exploring how your hormones are working without the influence of birth control hormones, you could experiment with coming off the pill for a couple of months. I'm not advising anyone to do this without seriously considering the impact of it – and definitely using alternative methods for protected/safe sex – but it is an option to think about. Without artificial hormones flooding your system, you have an opportunity to find your proper, optimal hormonal baseline with a DUTCH test. You can track your symptoms alongside your hormone levels, and that will show you what you're dealing with. We're all wired differently with hormone levels, and it's such a great thing to know what your levels are, and how you can then hack them ideally for you.

If you're older and heading into perimenopause, the trouble is, we aren't able to aim for our optimal hormone balance, because we never took the test when we were young enough to know what it was! Instead, we have to go on the received 'norms' and a trial and error approach until we find what works for us. But just be aware that we simply cannot understand our hormones and our symptoms accurately if we're still taking the pill.

What happens to our hormones during a regular twenty-eight-day menstrual cycle.

OK, I know loads of you might be thinking, *I frigging* wish *I had a twenty-eight-day cycle, I'm all over the place!* I hear you, but bear with me. This is just to explain what is going on hormonally in a 'regular' twenty-eight-day cycle so we understand what our biology is actually doing at different points in the month, and how our hormone levels change.

As with EVERYTHING hormonal, it's a complex dance between so many different hormones that can have multiple effects on us if they're out of balance, but the major ones in control here are oestrogen and progesterone.

The different phases of our cycle

Days 1–5: Menstrual

Or, in other words, your period! This is when you bleed, as your uterus sheds the lining that's built up in your previous month's cycle. Oestrogen and progesterone are both very low, so you won't be at your best.

Days 5–14: Follicular

After your period, your brain uses FSH to tell your follicles to start producing an egg. Oestrogen rises after about six days, which means lots of different effects: your uterus lining starts to builds up, you feel more energised and your mood improves (hurrah).

Day 14: Ovulation

You release an egg from your ovary, and your oestrogen levels peak.

Days 15–28: Luteal

The empty follicle breaks down and starts producing progesterone, which helps to thicken the uterine lining in case your egg is fertilised. Your levels of testosterone, oestrogen and progesterone begin to drop in the last part of this cycle, which can often trigger the famous PMT.

WTF IS PERIMENOPAUSE AND MENOPAUSE ALL ABOUT?

Like with EVERYTHING female-health related, the menopause used to be completely hushed up. Our mums and grandmas would whisper about this thing called 'the change' and when I was younger, I had no idea at all about what was in store for me – nor that it wasn't actually something to be ashamed of.

Medically speaking, the menopause is actually the time when you are no longer fertile. You are classified as having gone through the menopause once you've had a whole year without periods. But of course, this involves all your hormone levels drastically changing *again*, and like with everything hormonal, it doesn't happen in one smooth, simple way for everyone. Just as some women seem to sail through menopause without a hitch, many more deal with horrendous side effects from these fluctuating levels of hormones such as brain fog, insomnia, anger, rampant mood changes and hot flushes (to name a few!).

For many of us women, the perimenopause is actually a far tougher time than the actual menopause. Perimenopause – which is the transition phase leading up to menopause – can begin from anytime in our mid-thirties onwards, although the average women starts experiencing symptoms from their mid-forties. (But as we know, most of us are FAR from average!) The average length for perimenopause is four years, although it can last up to fifteen years in some cases. Flipping *heck*.

What happens to our hormones during perimenopause?

During the perimenopause, our hormone levels start to change dramatically. The levels of oestrogen and progesterone that we're used to having start to fluctuate from month to month. Basically, your ovaries can behave unpredictably: one minute they'll work and the next they won't. It's like going through puberty all over again – but this time, in reverse! This results in a rollercoaster ride with a whole host of life-altering symptoms.

Overall, this is what happens to our main female hormones during the perimenopause:

FSH levels go up and down (sometimes day to day) which causes . . .
Oestrogen to rise and fall

Progesterone stays lower, as we begin to ovulate less and more erratically.*

This will then have a knock-on effect on lots and lots of other hormones, as oestrogen receptors are all over our bodies. As I mentioned, I started researching perimenopause when I hit my late thirties and noticed how much worse my PMT was. Until I started looking into this, I was *completely* unaware that it could be perimenopause. I started tracking my cycle and mood on an app, and it slowly dawned on me – not only that I was starting to biohack, but that yes, I could be perimenopausal even though it was considered 'too early' by GP standards.

How to know if you're perimenopausal

So many women since time began have been dealing with the life-wrecking symptoms of perimenopause without a) knowing what's causing their symptoms or b) knowing what they can do about it. Even now, suicide rates are rising in women between the ages of forty-five and fifty-four, the prime age for perimenopause and menopause.† It's absolutely vital that we understand what's going on hormonally with us at this stage in life, and don't accept the misdiagnoses that are often (mistakenly) doled out by our GPs, who haven't even been trained properly in menopause!

One of the most common problems is GPs diagnosing women with depression or anxiety issues (and prescribing

* https://www.reproductivefacts.org/news-and-publications/patient-fact-sheets-and-booklets/documents/fact-sheets-and-info-booklets/the-menopausal-transition-perimenopause-what-is-it/
† https://www.itv.com/news/2021-11-16/suicide-rates-in-women-of-menopausal-age-rise

them antidepressants), whereas their problems are cycle and hormone related. I keep hearing so many stories like Pamela's on Instagram: 'I'm forty-two and have been suffering from brain fog, feeling overwhelmed and a whole host of other symptoms. I think I'm perimenopausal. Doctors think I'm being ridiculous.'

There's no actual test available to see if you're in perimenopause (I know, right?), so you have to go by your own metrics of how things have changed, which is why tracking is SO important. Look carefully at everything: have your usual period symptoms become more ferocious? Is your mood lower, or more erratic? Is your energy depleted? Are you having self-doubt and a negative internal monologue? And does it happen right the way through the month, or just at specific points? If it ebbs and flows it's going to be related to your cycle and so you need to be really aware of this and log it! You need to have as much information at your disposal as possible so you're able to go to your GP (if you want tests or HRT), and also hack your symptoms yourself.

The most powerful thing you can do is to be able to understand what's behind your various symptoms, what factors might be contributing towards them, and how you can change this! It's all about finding your pattern and using that knowledge as power.

A brief history of HRT

Thank GOD we are all talking about HRT more nowadays and the whole world is starting to understand how vital it is for millions of women. There's been a real societal shift in recent years around HRT, and rather than seeing it as a whispered 'nice to have' for older women (there's that shame about our biology again), women are becoming more vocal about how it literally helps them survive. I see it with my followers on Instagram a lot. Nikki says, 'My fuse had pretty much disappeared and I was losing my shit at everyone and everything but not realising how bad it had got. Started HRT three months ago and have noticed a big difference already.'

HRT was first made available in the 1960s, after scientists learned how to manufacture oestrogen synthetically. One crazy fact is that originally HRT was made from extracting oestrogen from the wee of pregnant horses! (The only brand that still uses this method today is Premarin, which is rarely prescribed in the UK.) HRT really took off big time in the 1990s, but there was a scare in the early 2000s when a study was published that linked HRT with higher rates of breast cancer, stroke and blood clots. The number of people taking HRT in the UK dropped during that time from 2 million to 1 million and I cannot imagine the amount of agony these women went through because they were too scared to take it.

Thankfully, the study that freaked everyone out was eventually debunked – it was shown that the increased risk was ONLY applicable to those who started HRT when they were seventy or over! For women who started HRT under the age of sixty – i.e. most of us – it was found that it had some protective health benefits instead.*

* https://www.womens-health-concern.org/help-and-advice/factsheets/hrt-the-history/

What's the difference between body-identical and bio-identical HRT?

Now this is a confusing one, and I've mixed up these before (it's easily done). Here's how to differentiate these two *very* similar-sounding sorts of HRT:

Bio-identical HRT (my personal choice) is derived from plant sources that have the same chemical hormones that our body has. They're custom-made for each individual and sometimes provide hormones in doses that aren't approved for women, such as DHEA (dehydroepiandrosterone). Bio-identical HRT is not available on the NHS because it's not regulated by the Medicines Regulatory Agency (MRA).

Body-identical HRT is HRT that is exactly the same as your body's natural hormones. It's usually in the form of oestrodial (the type of oestrogen we make during our fertile years) as a patch or gel, and also progesterone if needed. This is prescribed by the NHS and has been subject to safety testing.

I'd recommend doing some really intense, thorough research on practitioners if you want to explore bio-identical HRT. The NHS stuff is standardised so it may not be right for you, but like I've said a billion times before, we're all so different, there isn't one rule for all. I always recommend to my friends to explore HRT options if they're feeling horrendous. Look, we weren't supposed to be living this long so we need to boost our hormones as best we can before we all end up with broken bones from osteoporosis!

Over the years I've dipped in and out of taking HRT. I tried progesterone because I felt super anxious and I took a really low dose, but it just made me feel worse. I think I'm either progesterone-intolerant or need a higher dose. I have oestrogen gel too, that I smother on if I feel I need it, but it hasn't really worked that well for me (and I've covered why in the section below on my thyroid!). Before she died, my mum mentioned she had the same reaction to progesterone, which was really good to know. So much is down to genetics, so if you can, it's a good idea to ask your mum how she experienced perimenopause and menopause for a guideline as to how it might affect you.

MY THYROID REVELATION

As you all know, I'm on a constant journey of discovery myself when it comes to hormones and biohacking. I'm constantly researching new ways of treating myself, finding new experts to work with and digging down to what's really going on on a biochemical level. While writing this book, I started to feel really depleted: brain fog, low mood, low libido and weight gain around my middle that I couldn't shift with running or cryo.

All these symptoms looked like straightforward perimenopause, but in fact, I've discovered the root cause of it was my thyroid being completely out of whack. I found this out by working with a new expert, Justin Maguire, who had the most amazing and simple hack for finding out what was going on: I had to track my basal body temperature a few times a day.

Basically, our body temperature reflects our metabolism, which in turn, is determined by the hormones secreted by our thyroid gland. By logging my temperature (using an everyday thermometer), three hours after waking and then at three-hour intervals after that, and then plotting them on a chart, I was able to see a pattern emerge. I had a low body temperature (mine was 36.5 degrees, below 37 degrees is considered low), which means I have *hypothyroidism* – my thyroid isn't producing enough T3 and T4 hormones, which, as mentioned before, negatively affects the production of lots of other hormones. This could be a good place to start investigating.

This is just my own theory, but I believe a lot of my previous trauma is the reason my thyroid is knackered. My mum had a lot of serious health issues when I was little, which put me on high alert from a really young age. Me being me, I didn't process the trauma properly and instead, once I became a teenager, I dealt with it by constantly going out and favouring adrenaline and cortisol to energise me – putting my thyroid under massive strain. Once I got to my thirties, I think my body simply ran out of thyroid hormone and tipped me into hypothyroidism. That's why I drink gallons of coffee and run constantly to boost my dopamine!

Now, I'm treating my hypothyroidism with bio-idential thyroid grain, along with thyroid-supporting vitamins and minerals such as selenium, iodine, B12 and FORSKOLIN. It sounds gross, but it comes in capsules so they're easy to take. I've found a dose which works for me, and it has boosted my body temperature sufficiently so it's within a healthy range. I already feel more stable and able to cope with situations that would have overwhelmed me beforehand (such as dealing with loads of middle-aged drunk men at a golf tournament I was at with my sons recently!).

Like with all the advice in this book, I'm not saying this hack is a cure-all for everyone, and I've definitely got further work to do on myself when it comes to my baseline oestrogen and dopamine levels. But it's reminded me again that hormones are not a one-size-fits-all operation, and it's always worth exploring new options. Tracking your temperature – which you can do really easily yourself, and at home (just download a chart off the internet) – can reveal so much about your hormonal health.

HOW TO HACK YOUR HORMONAL CYCLE (ASIDE FROM HRT)

If you've come to this chapter first, all my advice will be completely new to you. However, if you've already read most of the book so far, this might seem a bit 'yeah, but I've read this already!' This is because SO much about our cycle relates to pretty much every other aspect of our lives: our mood, sleep, appetite, brain fog, etc.

As I've said, our hormones don't work in silos – which would have made them *so much easier* to write about, but science doesn't work like that! Instead, they are in a dance with each other all the time, causing a cascade of multiple effects and symptoms, and the root cause – e.g. a problematic cycle – can manifest very differently in all of us – e.g. poor sleep, heavy bleeding, PMT. Therefore, it's inevitable that a lot of what I say here will cross over with previous chapters.

As well as reading this chapter, I'd also recommend you gen up on the chapters that are relevant for your symptoms. So, if you've got raging PMT, you should also read the chapters on rage and low mood. If you crave carbs and sugar like a beast in the run-up to your cycle, you should also read the section on appetite and eating. And so on. But first . . .

The number one supplement I recommend – DIM

DIM (otherwise known by its snappy name, diindolylmethane) is one of most popular hormone-related supplements around. Our stomach acids naturally create DIM, an enzyme, when we break down cruciferous vegetables like broccoli, but you'd have to eat *bushels* of the stuff to make a difference this way, which is why it's much better taken as a supplement.

DIM is great because it balances out our oestrogen levels by breaking them down properly, which is why I recommended it in the chapter *Where Is This Rage Coming From?* when we covered oestrogen dominance. The research on DIM is limited so far, but it has been shown to detox excess oestrogen, favours production of protective oestrogen and reduces the destructive ones that are behind some cancers. Because of this, it reduces PMT and menopause symptoms. It regulates our energy and mood levels, and is well known for reducing acne, too.* A daily dosage is 200mg and I'd recommend giving it a try to tackle overall cycle-related hormonal issues.

Men need this too! DIM is excellent for men, who can also get oestrogen dominance – symptoms include man boobs, excessive belly fat, mood swings, irritability, low libido and depression.

* https://www.healthline.com/nutrition/dim-supplement#uses-benefits

HOW TO HACK YOUR HORMONAL CYCLE SYMPTOMS

Tackling PMT (pre-menstrual tension)

PMT is a beast, isn't it? In the days leading up to our period, our progesterone levels drop dramatically, which results in an imbalance of our sex hormones. During this timeframe, we're likely to experience some horrible symptoms – and depending on our unique makeup, what we experience will vary wildly. In fact, there are 150 recognised medical symptoms that women deal with in the lead-up to their period. *150!* They can include bloating, headaches and moodiness. Cramping, headaches and irritability. Fatigue, cravings (more on that below) and anger. *Delightful!*

I didn't suffer from PMT at all when I was younger, I've only noticed it in my older years since tracking it with the Flow app; and being able to do this has been great. It's only since that app came out that I could go, 'Oh my God – these are the days I'm more agitated and ragey.' However, the issue with PMT is that very often, the symptoms we suffer with are part of a bigger jigsaw puzzle – our symptoms, cycle and habits are all linked.

Because of this, it's sometimes hard to know where PMT ends and separate health issues begin. But there are plenty of ways we can tackle the hormonal imbalance of PMT and get ourselves into an optimal place throughout the month.

Get your pillars in place

By this I mean making sure you have your bedrocks of decent sleep, exercise, and enough exposure to daylight to boost your production of serotonin (the happy, cosy hormone). This in and of itself should be the first port of call. If you struggle with sleep, and don't get enough outside light – either because of your job, or lifestyle – read the *Why Can't I Sleep?* chapter for a deep dive into tackling this.

Boost fatty acids foods and EPAs

We already know that foods rich in omega-3 fatty acids are brilliant for SO MUCH, including promoting serotonin – helping our mood – and supporting our immune system and calming down our cortisol response. Overall, that's brilliant for helping us deal better with PMT. However, they've also been found to alleviate symptoms of PMT (hurrah!). One study that gave women omega-3 supplements found they had reduced levels of depression, anxiety, bloating, headache and tender breasts.*

Omega-3 fatty acids are found in fish, especially mackerel, salmon and oysters, plus nuts and other foods I list in the *Why Can't I Stop Eating?* chapter. You can also boost your intake with a supplement called EPA (eicosapentaenoic acid), which is found in omega-3 fatty acids. A good place to start is with a daily dose of between 1,000 and 2,000mg.

* https://pubmed.ncbi.nlm.nih.gov/23642943/

Take vitamins for progesterone

All the B vitamins are crucial here, because they're the building blocks that help us manufacture our reproductive hormones: B vitamins 2, 6 and 12 are key players in activating progesterone. One study actually showed that increasing the amount of B6 taken per day can raise progesterone enough to improve symptoms of PMT.*

We can get B vitamins in the food we eat – especially salmon, tuna, eggs and avocados – or alternatively, take a supplement.

Oestrogen detox with calcium D-glucarate

If you have PMT with sore boobs, cellulite, feeling moody and cranky, then you're more than likely not pooing out your oestrogen properly! Oestrogen needs to be detoxed to work well, otherwise you get a blockage in your liver and the wrong sort of oestrogen going round and round.

You can clear your detox pathways with DIM (see above) and also, please do read all the advice about oestrogen dominance in the section *Where Is This Rage Coming From?* Another great thing to take is calcium D-glucarate supplements, which help your liver function. Glucaric acid (the natural form of calcium D-glucarate) is a chemical in our guts that supports the liver, and is essential for ensuring that the excess oestrogen leaves our body. Without sufficient levels of calcium D-glucarate, hormones will be reabsorbed – hence, oestrogen dominance (which also affects men).

* https://www.letsgetchecked.com/articles/naturally-increase-low-progesterone-levels/

Sorting our irregular or heavy periods

Our thyroid gland is the master switch behind regulating our menstrual cycle, and when it's not working properly, our cycles can go completely crazy – making them too light, too heavy, or irregular.

As outlined earlier, if we have *hyperthyroidism* then we're producing too much thyroid hormone, which makes our periods light and short. With *hypothyroidism* we produce too little thyroid hormone and end up with heavy bleeding. In both cases, hypo and hyper, our cycles can end up completely irregular.*

Another classic root cause of irregular periods is perimenopause, as your sex hormones start to wildly fluctuate, especially as your oestrogen and progesterone levels deplete. My friend Becky knew she was perimenopausal when in addition to her periods going out of whack, she wanted to murder her husband for a few days each month!

Red light therapy to boost your thyroid

One great way to improve your thyroid function (without going on steroids, which can have all sorts of horrible side effects and become ineffective as the body adapts to them – hello, weight gain!) is with red light therapy. Red light boxes are now commonly used for balancing our hormones, and studies have now found that applying a low-level red light against our throat (where the thyroid gland sits) can positively affect our thyroid function!†

* https://www.verywellhealth.com/menstrual-problems-and-thyroid-disease-3231765

† https://rouge.care/blogs/rouge-red-light-therapy-blog/what-you-need-to-know-about-the-benefits-of-red-light-therapy-for-thyroid-health

This might seem like some kind of woo-woo magic, but it's actually down to the anti-inflammatory effect of red light therapy, which allows the thyroid to start working properly again. You can get a red light box for about £90, which is a bit pricey, but it's such a valuable piece of kit. It also works to boost collagen and improve skin conditions and sore muscles and joints, so it's a win–win.

Avoid xenoestrogens

Xenoestrogens are the 'bad' fake oestrogens that are found in loads of our daily plastics, cleaning products and even foodstuffs – you can find a proper list in the *Where Is This Rage Coming From?* chapter. They're a nightmare because they can interfere with normal female hormone activity, impair our thyroid function, and may increase your risk of breast cancer.

Try acupuncture

If you've had a baby, you might have been offered free acupuncture on the NHS in the run-up to giving birth – I remember having some sessions when I was having IVF, too. What's nuts is that acuptuncture's been around for thousands of years in Chinese medicine, the NHS offer it, but they don't actually know scientifically why it works – either for promoting fertility, preparing the body for labour or regulating your cycle. But it does!

In one study, a woman with severely heavy, irregular periods had ten sessions of acupuncture, which completely normalised her cycle.* Amazing, right? It's definitely something to look into having if you're dealing with irregular periods that aren't related to perimenopause.

Managing our crazy cravings

Now this is a biggie. It's something I still deal with (being at my core, a massive sugar addict), and so many of my followers get insane sugar cravings in the run-up to their period, in contrast to how they feel during the rest of the month. Lisa on Instagram says, 'I can eat really well until I'm due on and then the floodgates open. It's awful. I feel like I have no control, sometimes I don't even realise I'm back in the fridge or cupboard and it will be any food.' Sheila says very much the same: 'I can be so good all month, but the day before I come on my period and sometimes round ovulation I am literally ravenous. I have a meal and could eat the whole thing again . . . I just feel empty and unsatisfied and I end up bingeing then have low mood the next day . . .'

This happens (*of course*) because our feel-good hormones are dropping significantly as we approach our period, as are our sex hormones, which makes us hungrier, plus our cortisol (stress) levels are on the rise.† Therefore, it's no frigging surprise that we reach for the mint Aero, or whatever our own personal kryptonite is! But, as we all know, bingeing on chocolate and junk isn't a fix that makes us feel better in the long run. I'll share below how I adjust my eating around my cycle, but there are also some supplements you can take to dampen down those mad sugar cravings.

* https://www.ncbi.nlm.nih.gov/pmc/articles/PMC6088286/
† https://www.always.co.uk/en-gb/tips-and-advice-for-women/pms-and-menstrual-cramps/why-do-you-crave-chocolate-on-your-period/

Take NAC

NAC (N-Acetylcysteine) is now being touted as a great protocol for managing all sorts of cravings – as well as addictions – as it regulates the movement of glutamate around our nervous system. Glutamate is an amino acid essential for managing our learning and behaviours, and NAC can help regulate it.*

Add in chlorophyll

Does anyone remember learning about photosynthesis at school? Yes? Sort of? If so, you'll probably recall chlorophyll, which is a pigment present in all green plants. But it's actually an amazing nutrient for regulating blood sugar, and is now being promoted as a supplement to treat type 2 diabetes. It's amazing because it decreases our insulin spikes, which in turn, will reduce cravings.

Take L-glutamine

As I mentioned in the last book, I use an amino acid called L-glutamine in powder form to manage my sugar cravings. I put half a spoonful under the tongue (it tastes like talc!), which works wonders on me, and should get you over the first few days of craving. Obviously I chase this with a phat-fuelled coffee (coffee with MCT oil or keto powder). I don't need to take this all the time, but it's so good for balancing blood sugar and hitting that amino acid profile, which is what our appetite is searching for.

* https://www.ncbi.nlm.nih.gov/pmc/articles/PMC5993450/

EATING TO SUPPORT YOUR MENSTRUAL CYCLE

Days 1–14: Menstrual and follicular phase

While on your period and in the week afterwards, you're low on oestrogen and progesterone, but they will slowly begin to rise. This is the time to rest and restore, so drink plenty of chamomile tea, water, and support serotonin production with fermented foods like kimchi and good sourdough bread. While on your period, you'll be losing blood (obvs), so up your intake of iron-rich foods like red meat and seafood, as it contains heme iron – which is absorbed brilliantly by the body. On the flip-side, spinach (deemed a super food) contains non-heme iron which is not easily absorbed. More than 95 per cent of functional iron in the body is heme iron. Fun fact: clams contain more iron than beef liver! A few squares of good quality dark chocolate are great for replenishing your magnesium levels. I sometimes add dark cocoa powder to keto powder and collagen with water and stevia for a delicious drink.

Days 14–24: Luteal phase

At this point, your female hormones are at their highest level, so you're likely to be feeling at your best and most energised. Support your liver function and detox pathways with anti-inflammatory foods like vegetables, nuts, extra virgin olive oil, turmeric and oily fish (anti-inflammatory omega-3) as well as nutrient-rich meat, eggs and dairy.

Days 24–28: Pre-menstrual phase

As your levels of oestrogen and progesterone drop in the run-up to your period, you'll want to avoid foods that increase cramping, like caffeine and alcohol. Support serotonin production with leafy greens, turkey and chicken, red meat, bananas and cherries. This is the time when most of us fall prey to the sugar beast, so make sure you're nurturing youself with good foods – like fermented ones – that don't spike your blood sugar and leave you feeling worse afterwards. Magnesium-rich foods like avocado, bananas and nuts and seeds will boost your levels, which will stave off headaches and sleeping problems.

How I eat to support my cycle

I'm sharing this *not* because it's the blueprint for everyone else! There's no one 'ideal' way of eating, and hopefully throughout this book, I've banged on about this enough that you've got the message by now: you need to develop your own personal approach to hacking your hormones positively.

But it IS really beneficial to support your cycle as best you can by adjusting when and what you eat at different times in the month and making sure that the food you do eat is as ancestral as possible. By this I mean REAL food, not processed packaged crap, even if it is labelled as 'healthy' or 'sugar-free'. As we've seen over and over again, processed food is NOT going to help you feel better in any way – you need B vitamins, loads of iron and decent fatty acids to level out your mood.

A note on intermittent fasting

If you've read the other chapters in the book, you'll already know why I'm a huge advocate of intermittent fasting – which is essentially pushing back my eating window until later in the day. For my energy levels, and to stave off the sleepier serotonin/

melatonin hormone release, I've found that it really works for me. Also, I'm an addict by nature, and I know that if I start eating earlier in the day, I won't stop. My fatty coffee and essential amino acids that I add in (from the collagen powder) get me over that hump.

If you want to try this out for yourself, you don't need to follow my exact timetable, but should experiment with what works best for you. I'd suggest – as I did in the *Why Can't I Stop Eating?* chapter – that you don't make any drastic changes all at once. Just start by slowly shortening your 'eating window', i.e. the period in which you are eating, by eating breakfast a little later in the day and your dinner a little earlier. You shouldn't feel super hungry while intermittent fasting – I don't!

So, this is (generally) what I do and, importantly, WHY I eat like this. (Not included: falling off the wagon with a Chinese takeaway . . .)

Days 1–14: Menstrual and follicular phase

What's going on?

During this phase, I'm as kind to myself as possible by not revving up my sugar intake. It might be tempting, but I know how shit it will end up making me feel. Like I said just a moment ago, I'm an addict and know that if I start eating sugar, I won't stop, so moderation isn't an option for me.

Hormonally, I'm at my least resilient during these two weeks of my cycle. My oestrogen and progesterone levels have dropped right down, which has a knock-on effect on my dopamine, cortisol and every feel-good hormone I've got! My mood is definitely lowest during days 1, 2 and 3, as I've mentioned. So, I try to balance myself out and promote serotonin production by intermittent fasting, and eating fermented foods and those that contain tryptophan.

What I eat and when

On waking: Water with quality electrolytes
7 a.m.: Espresso with bovine collagen peptides
8 a.m.: Phat-fuelled coffee with MCT and collagen powder
1 p.m.: Bone broth, kombucha
3 p.m.: More bone broth with MCT oil, or celery juice if I'm bloated
4 p.m.: Protein shake with nut butter and nootropics
4.30 p.m.: Coffee with L-theanine
5.30 p.m.: Dinner options – spoonful of kimchi/sauerkraut to start, and then I'd choose from:

- Sourdough sandwich with organic meat and salad and Hunter & Gather (no veg oil) avocado mayo
- Homemade chicken or veg soup with bone broth base and salad
- Organic roast dinner
- Full English breakfast with sourdough
- Avocado and smoked salmon, scrambled egg and cheese on sourdough toast
- Dessert: chocolate sweetened with stevia, or home-made cocoa avocado mousse

9 p.m.–11 p.m.: Snacks:

- Olive oil crisps (some Torres flavours are olive oil only – but always check the label!)
- Cheeseboard with sourdough crackers

- Homemade chicken liver pate on sourdough toast with butter
- Kombucha

10 p.m.: Chamomile tea

Days 14–24: Luteal phase

What's going on?
At this stage in my cycle, I don't need to make as much serotonin because I have plenty of oestrogen – I'm manufacturing the feel-good hormones like a beast! It's the time of the month when I feel most energised, creative and interested in life. I cut down on carbs more here as I'm not craving them as much.

What I eat and when

On waking: Water with quality electrolytes
7 a.m.: Espresso with bovine collagen peptides
8 a.m.: Phat-fuelled coffee with MCT and collagen powder
12 p.m.: Phat-fuelled coffee with MCT and collagen powder
1 p.m.: Kombucha
3 p.m.: Phat-fuelled matcha tea with MCT powder and L-theanine
5 p.m.: Bone broth shot with MCT oil
6 p.m.: Dinner options as the menstrual and follicular phases but I'll have fewer carbs, so I'll cut the sourdough in half
9 p.m.–11 p.m.: Snacks

- Nut butter on an apple
- Cheese and celery, with fruit, on sourdough crackers with grass-fed butter

- Activated nuts baked in honey and sea salt

10 p.m.: Chamomile tea

Days 24–28: Pre-menstrual phase

What's going on?

In the days leading up to my period, I'm definitely at my hungriest! Because of this, I'll eat earlier on in the day, and listen to what my body needs. I also feel a bit more aggressive, which for me is OK, as I just incorporate a bit more running into my schedule, and cut down on my caffeine intake, which can tip me into more aggression. (However, Matthew says I'm nowhere near as bad with the blind rage as I used to be! I'm just a bit more snappy than normal.)

What I eat and when

On waking: Water with quality electrolytes
7 a.m.: Chamomile tea with MCT powder (though sometimes I do have coffee)
8 a.m.: Kombucha
9 a.m.: Full English breakfast with sourdough and grass-fed butter OR smoked salmon, avocado and eggs (I'll also take digestive enzymes containing amylase [breaks down carbs], protease [breaks down proteins] and lipase [breaks down fats] to reduce fatigue during digestion)
12 p.m.: Protein shake with bone broth powder mixed with raw cow's milk or coconut milk, one banana and nut butter
Stevia sweetened chocolate and kombucha throughout the day
6 p.m.: Dinner with the family from the same list as the menstrual and follicular phases. If we have a take-away I always take my own brand Rise and Shine tablets and drink a

cleansing tea afterwards. I allow myself to indulge a bit more around this time of the month, but I don't see it as a failure or a reward, it's just a normal hormonal reaction. I also try to do a few niacin saunas to help me detox (see my previous book *It's Not a Diet* for this protocol) and give my mood a lift, and also go for a walk alone to calm my brain and reduce inflammation.
9 p.m.–11 p.m.: Snacks

- Olive oil crisps (some Torres flavours are olive oil only – but always check the label!)
- Cheeseboard with sourdough crackers with butter, fruit and/or honey
- Homemade chicken liver pate on sourdough toast with butter

10.p.m: Chamomile tea

GO BACK TO BASICS

I really believe the key to getting hold of your menstrual cycle – whatever your issue – boils down to two essential things. Firstly, TRACK your cycle and symptoms to find out what is happening to you. Secondly, WORK WITH your cycle, not against it! By this I mean once you understand where you're at in your cycle, support it as best you can with what you eat and what you do. You want to thrive as best you can throughout the month, and not be at the mercy of your plummeting hormones!

So, take control, take the power into your hands, and get back to basics.

My cycle now

Cycle length ___ days	*Energy levels*	*Symptoms*	*Cravings*	*Mood changes*
During my period				
Follicular phase				
Ovulation				
Luteal phase				

How I'm going to support my cycle

Cycle length ___ days	*Eating plan*	*Supplements*	*Exercise*	*Mood changes*
During my period				
Follicular phase				
Ovulation				
Luteal phase				

DR E – HOW TO TALK TO YOUR DOCTOR ABOUT YOUR CYCLE

Women are increasingly seeking out integrative health as an approach to better understanding themselves and their hormonal cycles. While I fully support taking a holistic mind–body approach to your menstrual cycle it's still important to screen for underlying disease or pathology when it comes to having symptoms.

I would advise you to seek professional medical advice if you are experiencing any of the following symptoms:

- Irregular bleeding, including spotting between periods, or after menopause
- Irregular vaginal discharge
- Post-coital bleeding (after sexual intercourse)
- Breast changes – irregular appearance to the skin, lumps or pain
- Weight gain, increased hair growth and irregular periods – think polycystic ovaries
- Vaginal dryness, itching or irritation
- Menopausal symptoms (it can start earlier than forty-five, and is called Premature Ovarian Insufficiency in those under forty years of age):
    - Hot flushes/night sweats
    - Mood impairment – anxiety, mood swings, low mood
    - Recurrent urinary tract infections/vaginal dryness/irritation
    - Loss of sexual desire and libido
    - Sleep disturbance

    - Fatigue, aches, headaches can all be related
- Amennorhea (when your periods stop, or don't start at all)
    - Primary: no menstruation before thirteen years old and no secondary sexual characteristics (such as breast development). No menstruation by fifteen years old with normal secondary sexual characteristics
    - Secondary: no period for three to six months after normal periods, or six to twelve month absence after irregular periods

Long term use of the pill can have an effect on your hormonal cycle and, if you come off the pill after a prolonged time, it's unusual for your cycle to return to normal straight away. It may take a while for you to start ovulating again. After a long time on the pill, it is common to experience nutrient deficiency and gut microbiome changes. This means that in order to return to hormonal balance you may need to address your gut, as well as detoxification support for the liver; and finally look at your other hormones as a base too (adrenals and thyroid).

Doctors are seeing increasing numbers of women seeking to become pregnant later in life, with fertility treatments the promised land of hope. Fertility treatment does work, but it doesn't work for everyone. There is more to becoming pregnant than simply taking hormones to stimulate ovarian growth and proliferation. My advice is to seek to create balance in your biology in your preparation for pregnancy – with nutrients, diet and lifestyle changes: the integrative approach.

During fertility treatments, we've seen success with supporting ovulation and peri implantation with supportive treatments like Hyperbaric Oxygen Therapy and NAD+. There are studies to show the increased success rate of fertility therapy using these simple supportive measures in a smart way.

HRT and BHRT are increasingly being used and better understood by GPs across the country. I recommend speaking to your GP about the risks and potential benefits of hormonal replacement therapy. There are various forms of delivery: patches, creams, gels, oral tablets, vaginal pessaries. The other benefit of HRT is reduced fragility fractures in bones. I suggest you have regular reviews of your HRT prescription – every three months is advisable.

Keep a symptom diary! Track your symptoms with apps and use these records to help make the most of your time with your GP.

Medical testing:

(suggested not exhaustive biomarkers – these tests will be personalised by your practitioner after consultation)

FSH, oestrodial, progestoerone, testosterone

Thyroid baseline + adrenals baseline: TSH, T4, T3, morning DHEA and cortisol

Blood sugar response: HbA1c, fasting insulin and fasting glucose

Full blood count

Liver function test

Integrative testing:

(suggested not exhaustive biomarkers – these tests will be personalised by your practitioner after consultation)

Homocysteine, B12, folate – looking for methylation issues

MicroNutrient status: Mg, Zinc, Cu+, etc. – organic acid test is useful, or Genova NutrEval

Gut health: microbiome, leaky gut, parasites, funghi, etc. – comprehensive stool test

DUTCH Complete or Cycle mapping

(there are others but these are the important ones to start with)

Tech:

Keep an eye out for Femtech – technology tools to support women's health.

Flo.health and Clue – cycle trackers

Health wearables such as Ōura use body temp to track your cycle.

A WORD BEFORE I GO . . .

It feels strange for me to write a conclusion to this book, because I don't feel that I'm ever 'done' when it comes to knowledge about my body, and the incredibly complex hormonal dance that dictates our wellbeing. Even while writing this book, I've found out more and more information, and scientists are discovering and publishing new insights all the time. So I don't think any of us can ever think, OK, right I'm done now! New research comes out all the time, and who knows what's out there that we don't know about yet?

Having said that, my goal with this book has been to pack it full of as much useful, tangible knowledge as possible that *you can use*. And I wanted to make it easy to relate to, not full of confusing jargon. That's why this book is symptom-led, rather than hormone-led. We all live in the real world, we want to know what's behind the symptoms that affect us – and most importantly, what we can do to help ourselves.

In reading this book, I hope you've understood that the power to feel like your optimal self is truly in your hands. You don't have to put up with feeling exhausted, or ragey or so overwhelmed you struggle to function. With the guidance on these pages, and Dr E's advice, it's my goal for you to feel equipped to interpret your symptoms, see what's missing and try out new approaches. You have all the tools at your disposal to hack your hormones positively and THRIVE. Your symptoms are the most important piece of artillery you have – it means your body is telling you something. And don't worry – like I've said before – if your GP says you're 'normal' yet you don't feel it. You know yourself better than anyone, and you can make positive changes through these hormonal hacks.

Hacking your hormones isn't a one-size-fits-all solution. If you haven't already done so by this point, please fill in the charts, try things out and also enjoy the process! Take pleasure in digging around and finding out what your body responds to.

There are so many benefits to be found on your journey of discovery, not only for you but for the people close to you. When you find things that make you feel better and more energised, share them, pass them on. I've felt so good sharing my journey with my followers on Instagram over the years, and I'm not stopping now! I'm experimenting all the time, so if you don't already, follow me @daviniataylor. It's a lifelong journey of learning for all of us, and it's so much better when we do it together.

Davinia x

APPENDIX 1

GLOSSARY

5-HTP

5-Hydroxytryptophan (5-HTP) is an amino acid that your body naturally produces. Your body uses it to produce **serotonin** after converting it from **tryptophan**.

Acetylcholine

Based in the central nervous system, acetylcholine is the most common neurotransmitter we have. It's in charge of muscle control, memory and sensations. Because it regulates our brain speed, if we have low acetylcholine levels we might have memory problems, difficulties in learning and thinking creatively.

Adaptogens

Herbal remedies that come from plants, like ashwaghanda and lingzhi.

Adrenaline

Adrenaline is our 'fight-or-flight' stress hormone, regulated by the adrenal glands which are just above each of our kidneys. It's there to keep us safe by giving us a surge of energy when our brains perceive a threat, and we feel that jittery, nervous sensation.

Amino acid

Molecules that combine to form proteins, sometimes called the 'building blocks of life'.

ATP

Otherwise known as Adenosine 5'-triphosphate by people who don't get tongue-tied, ATP is in every one of our cells acting as its 'petrol' by transferring energy.

Central nervous system

Aka our body's command centre, the central nervous system controls everything from our movements and thoughts to automatic processes like digestion and breathing.

Cholesterol

A waxy substance found in every cell in our body. Cholesterol is vital for producing hormones as well as making liver acids that absorb excess fat during our digestive process.

Collagen

A long chain of linked amino acids which provide our bodies with the structure they need to work. After water, collagen is the most abundant substance in our bodies, but it depletes from our early twenties onwards.

Cortisol

Famous as the 'stress hormone' and produced in our adrenal glands in a big cortisol pulse once a day, cortisol also plays a vital role in many other functions. It's an alpha hormone, meaning that it's one our bodies will prioritise producing over many others. However, cortisol levels can easily become too elevated, leaving us stressed, anxious and fearful.

Cytokines

These are peptides (small proteins) that are produced by our cells and regulate various inflammatory responses in our immune system.

Digestive enzymes

The pancreas makes three main digestive enzymes which break down foods. Without them, your body can't break foods down so that nutrients can be fully absorbed. Amylase breaks down complex carbs and is also made in the mouth. Lipase breaks down fats. Protease breaks down proteins.

Dopamine

The master hormone behind pleasure and pain. Dopamine is a fascinating one, because it plays a major role in regulating sleep, appetite, brain fog, focus, and overwhelm – as well as being the main hormone behind addictive behaviours. It's produced in the brain, and drives our reward pathways, motivating us to achieve outside of ourselves. When it's low we can feel down, hopeless, and just completely 'meh' about everything.

Emulsifiers

Chemical additives in our food which bind ingredients like oil and water together to make things appear creamy. They can inhibit the gut's ability to detect what's in our food and therefore mess with our leptin response.

Endocrine system

The network of glands that make our hormones. It includes the hypothalamus, pineal gland, pituitary gland, thyroid gland, parathyroid glands, thymus, adrenal gland and pancreas.

Endorphins

Often called our 'feel-good' hormones, endorphins help us deal with pain and also feel pleasure. They're released after activities like exercise, sex and laughter.

FSH

FSH (follicle stimulating hormone) is produced by the pituitary gland and has a key role to play in our menstrual cycle. Every month it causes an egg to mature in the ovary and also stimulates the ovaries to release oestrogen.

GABA

GABA (gamma-aminobutyric acid) is an anxiety-inhibiting neurotransmitter that slows down our whirring brains and promotes a calm, relaxed feeling. It's produced in the brain, and its main function is to regulate our immune response, controlling our fear and anxiety when our neurons become overexcited.

Ghrein

Our 'hunger hormone', ghrelin is mainly produced in the stomach. Its primary role is to regulate our appetite by sending a signal to our brain that it's time to eat. Levels vary throughout the day and can be triggered by things like smells, regular meal times and drops in blood sugar.

Hormones

Our body's chemical messengers. They're created by special cells in our endocrine glands and are released into our bloodstream to send a message to another part of our body.

Insulin

Insulin regulates our blood sugar levels. It's created in our pancreas and is released when our body breaks down food into glucose (sugar), which our cells use as energy. As the glucose moves around our cells, it triggers an insulin response which then brings our blood sugar levels down to normal levels. Eating too many processed, sugar-spiking foods is highly detrimental for our insulin levels and can increase risk of developing type-2 diabetes.

Leptin

The hormone that controls our satiety – i.e. the sensation of being full. It's produced by our fat cells, which send a signal to our hypothalamus that it's time to stop eating. Many processed foods now mess with our leptin receptors, which means we find it harder to manage our appetite as we're not receiving the satiety signals.

LH

Luteinising hormone (LH) is made in the pituitary gland and stimulates the release of progesterone, as well as causing the mature egg to be released by our ovary.

Melatonin

Best known as our sleep hormone, our bodies release it about two hours before we go to sleep. It's produced by our pineal gland and is the final stage of the tryptophan–5-HTP–serotonin–melatonin sequence that allows us to fall asleep. Melatonin levels can be impacted by light, temperature and what we eat.

Mitochondria

Sometimes called the 'energy factory' of our body, we have thousands of mitochondria in almost every cell. They break down food molecules and produce **ATP**, as well as perform lots of other essential functions.

Neurotransmitters

Brain signals that impact our thoughts, feelings and automatic responses like movement and heartbeat. Many hormones, like serotonin and dopamine, also work as neurotransmitters.

Nootropics

Compounds that can 'turn the mind' (from Greek, *nous trepein*), providing safe cognitive benefits which include enhanced memory and learning ability, a stimulant or sedative effect, protecting the brain or helping your brain to function under stress. Nootropics allow you greater control over your mood and brain energy.

Noradrenaline (OR norepinephrine)

So good they named it twice, this is adrenaline in the brain, and is important for our alertness, energy levels and brain function. It works in tandem with dopamine.

Oestrogen

One of the two central female sex hormones, oestrogen is responsible for developing and maintaining our reproductive system. It's produced in the ovaries pre-menopause and by the adrenal glands in our body fat post-menopause. Fluctuating oestrogen levels, whether too low or too high, can have multiple effects on us, as we have oestrogen receptors all over our bodies.

Oxytocin

Aka the 'love hormone', because it gives us warm, cuddly, fuzzy feelings. It's produced by the hypothalamus in our brains, and controls everything from sexual arousal to trust, romantic feelings and bonding, especially between mothers and babies. It's the hormone of social bonds.

Parasympathetic nervous system

Our 'rest and digest' system, which regulates all the automatic functions our body performs when at rest, like digestion. It puts our body into its restorative, calm mode, slowing down our heart- and breathing rate, as well as contracting our bladder.

PCOS (Polycystic Ovary Syndrome)

Facial hair, irregular periods, anxiety, PMS, miscarriage, and more! Look into low-carb (keto) or carnivore diets and implement anti-inflammatory protocols (see *It's Not a Diet* book).

Progesterone

One of the two major female sex hormones, it's released by our ovaries during the middle of our cycle and gets our uterus ready each month for a fertilised egg. When progesterone levels are high, we feel a calm, nurturing sense of wellbeing. It works in complete synthesis with **oestrogen**.

Serotonin

Aka our 'happy hormone', healthy serotonin levels are essential for making us feel good, cosy and safe. It works as both a neurotransmitter in our brain and as a hormone through our bloodstream, and 90–95 per cent is manufactured in our gut. When we have low serotonin, we don't feel joyful or safe and it can often lead to depression and other problems, like insomnia.

Sympathetic nervous system

Working alongside our **parasympathetic nervous system,** this one controls our 'fight-or-flight' response. It gets us ready for action by arousing bodily actions like sweating, increasing our heart rate and alertness.

Testosterone

The most abundant hormone in women, testosterone makes effort feel good! It's produced by the ovaries and adrenal glands and helps us build muscle, burn fat, and keep our energy levels and libido healthy. Our testosterone levels also interact with **cortisol** and **oestrogen**.

Thyroid

Our thyroid gland is situated in our neck and produces two hormones that regulate our bodies' energy levels (metabolism) and our menstrual cycle, T4 (thyroxine) and T3 (triiodothyronine). It's very common to have thyroid disorders where your body produces too little or too much thyroxine, which can impact your levels of other hormones like **progesterone**.

Tryptophan

An essential amino acid that we need for growth and lots of other metabolic functions. We get it from food or supplements, and the body converts it to 5-HTP, then **serotonin**, then **melatonin**.

Vagus nerve

This is the longest nerve in the body, running from our brain down to our gut, and it's the main communication highway of our **parasympathetic nervous system** – i.e. our rest and digest system.

Vitamin D

Or rather, hormone D, as that's what it actually is (as if you haven't heard me bang on about that enough by now). It's produced in the skin, in response to sunlight, and helps absorb calcium from the gut into the bloodstream as well as helping regulate sleep.

Xenoestrogens

These are synthetic industrial chemicals found in various plastics, preservatives and pesticides. They can act as oestrogen or even block normal oestrogenic activity by acting as an anti-oestrogen. Watch out for food and drink in plastic containers.

APPENDIX II

RECOMMENDED SHOPPING LIST: MY TOP FIVE BUYS BY SYMPTOM

Why Can't I Sleep?

Blackout blinds

An absolute essential for making your bedroom sleep-ready – even small amounts of light on the skin can trigger cortisol, which is horrendous for drifting off. You can get your curtains lined with blackout fabric, buy made-to-measure blinds, or even a portable blackout sheet with suction cups online from as little as £20.

Blue-light-blocking glasses

These help promote sleep by stopping the blue light waves from our phone, tablet and TV screens that trick our brains into thinking it's daytime – and therefore inhibiting our melatonin release. You can buy them online now for as little as a tenner.

Fermented foods: kimchi, kombucha and kefir

Our gut microbiome needs to be super-healthy in order to produce enough serotonin to then become lovely, sleepy melatonin. To do this it needs a diverse gut microbiome, and you can support this with fermented foods. They're really easy to get now, so whether you prefer kefir yogurt, slightly fizzy kombucha, fermented vegetables in kimchi, add it in.

Magnesium

Such a brilliant supplement for regulating sleep, magnesium activates the part of our nervous system that gets us calm and relaxed, plus it activates our GABA receptors which calm us down. Magnesium glycinate is the most popular supplement for sleep, and is easier on the gut than many other forms of magnesium (which are sometimes used as laxatives!). A good daily dose is 250–300mg. I also take magnesium threonate for my brain and citrate for constipation.

Zinc

Healthy levels of zinc have been found to reduce the time it takes you to fall asleep and increase the overall amount of time you spend asleep, too. A daily dose can be between 15 and 25mg.

Why Can't I Stop Eating?

Apple cider vinegar

A spoonful of apple cider vinegar before you eat (or diluted in some water) will kick off your digestive enzymes, slowing the breakdown of carbs in your gut. It also dampens down your insulin response, managing your blood glucose levels better and so helping to regulate your appetite.

Berberine

Berberine is a bioactive compound derived from tree bark and has been shown in studies to reduce the secretion of leptin, our appetite hormone. It's a completely safe, natural supplement that reduces our blood sugar levels and brings the insulin response right down. You can get it in tablet form to take before you eat.

Collagen peptide powder

Collagen contains amino acids which is what the body is craving when we're hungry. Adding a scoop of quality bovine collagen powder to your morning fatty coffee (see below), will help guard against the cravings for toast and cereal.

L-glutamine

This is an amino acid that I used personally when I was weaning myself off sugar. You put a spoonful of this powder under your tongue for thirty seconds, it gets into your bloodstream and it dampens down the sugar craving and stops you reaching for the Bounty/Mars/Dairy Milk/delete as applicable.

MCT oil/powder

MCTs (medium-chain triglycerides) contain amazing fatty acids derived from coconut oil. They're quickly absorbed into the bloodstream and sate your appetite because they contain good fats and get your brain and body switched on. Add some to your morning coffee to manage cravings, along with the collagen.

Why Does It Feel Like I'm Losing My Mind?

Choline

We can boost our body's production of acetylcholine, which helps us focus and learn, with certain foods like eggs, broccoli and liver, but also with a choline supplement. A daily dose can be between 100 and 500mg. An egg delivers 147mg.

L-theanine

L-theanine is an amino acid, isolated from green tea, that's great for dampening down panic and promoting calm. I take it with my coffee and just before bed. Daily doses can be between 200mg and 400mg.

MCT oil/powder

MCT (medium chain triglycerides) fat found in coconut oil is vital for supporting healthy hormone production. MCT is involved with the production of ATP – our cells' 'petrol' – which improves alertness, memory and mood. Research also shows that MCT increases antioxidant levels in the brain as well as serotonin, which has an anti-stress effect. I have MCT keto powder in my morning fatty coffee, and it's had a transformative effect on my mood and focus.

Mucuna pruriens (L-Dopa)

This is a natural adaptogen that has its basis in Ayurvedic medicine. It has high levels of a dopamine pre-cursor called L-Dopa, so supports the body's production of this fantastic hormone. A daily dose can be between 15 and 30mg.

Valerian root

This natural extract has been found to increase the levels of GABA in the brain, lessening anxiety. As it's also used as a herbal remedy for sleep, if you're just taking it for focus you should take a smaller dose – around 120 to 200mg three times a day – to avoid feeling sleepy.

Why Do I Feel So Low?

EPA

My top supplement for low mood; EPA (eicosapentaenoic acid) is found in foods rich in omega-3 fatty acids. EPA has been found to be hugely beneficial, as it not only reduces inflammation but can have the same effect as antidepressants. A recommended dose is around 2,000–4,000mg per day.

Epsom salts

Epsom salts are actually a type of magnesium (magnesium sulphate) – one of those super minerals that is essential for so many functions. Taking an Epsom salts bath can really help regulate our cortisol levels and increase our calming GABA neurotransmitter levels.

Ginseng

Ginseng, a type of Panax root, has been used for centuries globally as a powerful nutritional supplement. It has an anti-stress and anti-inflammatory effect and helps relieve fatigue. A typical daily dose is between 200 and 400mg.

Kimchi, kombucha and kefir (again)

Fermented foods are incredible for boosting your healthy gut bacteria, which will support serotonin production. Stack your diet with your preferred fermented mood-boosting foods, whether that's kimchi (pickled vegetables), kombucha (a type of fizzy green tea) or kefir (fermented yogurt).

Rhodiola rosea

A herbal adaptogen used worldwide to tackle low mood. It has a marked effect on stress-related fatigue and depression as it positively influences levels of serotonin and dopamine in the brain. You can take 200mg once or twice a day.

Where Is This Rage Coming From?

Ashwaghanda

This is an amazing, relaxing adaptogen that has been found in many studies to actually reduce cortisol and can help mitigate our low light levels, too. What's more, it can help you sleep better. As mentioned before, it can cause anhedonia, aka feeling meh! So be aware of that and pay attention to how you feel if you try it.

L-theanine

Managing your cortisol levels is vital to stop feeling so ragey and hunted. Taking L-theanine, an amino acid, each day (with or without your coffee, depending on how sensitive you are to caffeine), is fantastic for blunting the stress response. Daily doses can be between 200mg and 400mg.

Red light device or box

Bright, natural light in the mornings stimulates our cortisol response at the right time. If we don't get it stimulated then, it can then run rampage when we don't want it. Hack the low winter light levels by using a red light box in the mornings if the sun isn't out.

Sauna pod

Getting hot and sweaty is a great way to support your oestrogen detox. You don't need to get all lumberjack-y and build your own wooden sauna cabin as now you can buy your own sauna suit or portable pop-up sauna pod to use at home. Prices vary, as I'm sure you can imagine! It's another investment purchase, but one well worth it, in my opinion.

Turmeric

This spice helps detox our oestrogen, and it's anti-inflammatory too. You can either mix it with a small bit of water in the morning to have as a turmeric shot or sprinkle a little on whatever you're eating for lunch. Don't forget to combine it with black pepper to make it more bioavailable.

WTF Is Up With My Hormonal Cycle?

B2, B6 and B12 vitamins

These three B vitamins are essential for helping us activate oestrogen and maintain healthy progesterone levels, which can reduce PMT symptoms. You can supplement your diet with 100mg of B2 and B6 per day, and 500μg of B12. (The 'μ' symbol means one thousandth of a milligram, so just beware!) Opt for P-5-P (pyridoxal-5-phosphate), the active form of B6. A cofactor for over 100 enzymes.

Calcium D-glucarate

Another fab supplement for helping us detox our oestrogen properly, calcium D-glucarate helps support our liver function. You can start taking 200mg twice a day and gradually increase to 500mg.

DIM

DIM (diindolylmethane) is an enzyme our bodies need to balance out our oestrogen levels by breaking them down properly. It can help reduce PMT and menopause symptoms as well as balance energy and mood. A recommended daily dose is 200mg.

NAC

NAC (N-Acetylcysteine) is another amino acid that is great for regulating our cravings – perfect in the run-up to our periods. A general daily dose is between 600 and 1,200 mg. It is a precursor to glutathione, our master anti-oxidant.

Red light therapy box or device

Applying red light directly to our thyroid gland (at the base of our throat) has been found to support thyroid function, which regulates our cycle. You can spend anything from about £40 to £200 on a device, which sounds pricey but it should last for years. I'd recommend doing your own research and reading reviews to see which one would suit you as there's loads of choice.

APPENDIX III

RECOMMENDED SUPPLIERS

I'm often asked on Instagram for specific supplier recommendations and, while these are changing all the time, I thought it would be a good idea to list some of the places where I do my own health and biohacking shopping.

I can't mention supplements and suppliers without giving a shoutout to my own business, WillPowders! I started WillPowders in 2021 with the aim of bringing together the best ingredients to produce the most practical and bioavailable supplements on the market, and I'm so proud of our products. I remain constantly inquisitive and am always evolving the range, so checkout www.willpowders.com for products that have been developed specifically to target the hormonal issues we might be facing. Using all the information I have learned about supplements and working with my biochemist and endocrinologist, I have created a one-stop shop for some of the most common hormone imbalance symptoms I get asked about.

In addition to my own business there are so many others out there that I believe in, have bought from and find inspirational in the fight against chronic disease and misery. Some of my favourite suppliers are listed below and I hope that these resources will help you identify what's right for you and your lifestyle choices.

online food services

planetorganic.co.uk
wholefoods.co.uk
welleasy.co.uk
weareheylo.co.uk
fieldandflower.co.uk
oddbox.co.uk
gazegillorganics.co.uk
ableandcole.co.uk
eversfieldorganic.co.uk
coombefarmorganic.co.uk
thefishsociety.co.uk
coast-to-home.co.uk
riverford.co.uk
fishforthought.co.uk
shop.rickstein.com

raw dairy

naturaler.co.uk – for local dairy farms near you

fermented food and drink

lovingfoods.co.uk
myfermentedfood.com
thesourdoughco.com
modernbaker.com
jasonssourdough.co.uk

bottlebrushferments.com
labrewery.co.uk
kombuchawarehouse.com

protein powders

Brilliance Broth by willpowders.com
Dr Axe Ancient Nutrition at iherb.com
planetpaleo.com
Ancient Nutrition at luckyvitamin.com
Vegan Low-Carb Protein at gardenoflife.co.uk
organicwhey.com

collagen

willpowders.com (Grass-Fed Bovine Collagen Peptides)
hunterandgatherfoods.com (Grass-Fed Bovine Collagen Peptides)
ossaorganic.com (Grass-Fed Bovine Collagen Peptides)
Vegan Naked Collagyn at ancientandbrave.com

bone broth

Brilliance Broth by willpowders.com
planetpaleo.com
drgusnutrition.co.uk

electrolytes

Electrotide – willpowders.com
e-lyte – bodybio.co.uk

nootropics

Pure C8 MCT Oil – willpowders.com
Pure C8 MCT Oil – hunterandgatherfoods.co.uk

Nootropics Range – willpowder.com
troscriptions.com
onnorlife.com
dirteaworld.com

supplements

Hormone Hacker range – willpowders.com
Designs for Health range – supplementhub.co.uk
thorne.com
highernature.com
veridianhealth.com
pippacampbellhealth.com
hum2n.com
drsturm.com

testing

omnos.me
thriva.co
lifecodegx.com
medichecks.com
letsgetchecked.co.uk
freestylelibre.co.uk – continuous glucose monitor
joinzoe.com

equipment

redlightrising.com – red lights
functionalself.co.uk – general biohacking store
aquatrue.co.uk – reverse osmosis water system
waterfilters.co.uk – Berkey water filters
ouraring.com – health tracker
circular.xyz – health tracker

whoop.com – health tracker
firzone.co.uk – infrared pop up sauna
infraredsaunas.co.uk – infrared sauna for 2+ people
brassmonkey.co.uk – icebath
dryrobe.com – changing robe
voited.co.uk – changing robe
ikea.com – chill pad
eightsleep.com – sleep fitness technology
Lumen.me – helps to improve your metabolic flexibility
hackswellbeing.com – biohacking equipment

apps

Flo health – symptom tracker
mysysters.com – perimenopause symptom tracker
Wim Hof – breathwork and cold exposure
HeadSpace – guided meditation
MindValley – self improvement

practitioners

Dr Enayat – hum2n.com
Justin Maguire – autonomiccoaching.com
Pippa Campbell – pippacampbellhealth.com
Rosemary Ferguson – rosemaryferguson.com
Dr Tamsin Lewis – wellgevity.com
Gary Brecka – garybrecka.com
Dr Barbara Sturm – drsturm.com
Dr AJ Sturnham – thedecree.com
Inge Theron – facegym.com
omnos.me

ACKNOWLEDGEMENTS

This book is dedicated to all the women whose hormonal imbalances have seen them subjected to the most abhorrent medical interventions and social shame in the past. The embarrassment, the misunderstanding and the neglect of women's health are finally being addressed, and not a moment too soon. We need choice and real investigation into our hormonal complexities, not just what suits the profits of pharmaceutical companies. I hope this book can be a part of that long-overdue cultural change.

So on that note, credit and thanks to my incredible manager, Becca Barr, who has tirelessly supported me and believes in my message for better health for all. You have truly helped turn my life around and therefore subsequently helped thousands of others to feel 'healthy and happy' for the first time in decades. Thank you for taking a chance on me and some of my 'out there' ideas! You are brave beyond words.

A big thank you to Becky Howard, who worked so brilliantly with me in getting all of my ideas down on the page, and the whole team at Orion who have helped to bring this new book into the world. Thanks, too, to Alan Strutt and Gemma Sheppard for the great cover shoot in Marbs. Teresa Paul PR, I love your brilliant mind and infectious personality.

I also wanted to thank all my followers on the 'gram' (@daviniataylor) who support me in my ups and downs, highs and lows and all the way round adventures through this thing we call life. Thank you. You are family.

Of course I want to thank my actual family, my adorable Matthew and my beautiful boys, who help me be open and honest about hormones and how they affect us all. When we address our hormonal imbalances we also help our families have a better understanding of overall physical and mental health. As the mother of boys I think it's my duty to help my boys understand women are a force of nature to be respected and ultimately partnered with on an equal footing.

Matthew, I love how you love me, your dedication to me and my crazy adventures is the reason why I can put pen to paper, I know you've got my back. I love you so much words cannot explain.

And I have to thank the gang in the office at my business, WillPowders. Thank you, Sarah Lundy and Gill Cefai. I love our chats. I love brainstorming with you, doing the been-there-done-that-got-the-T-shirt-and-still-smiling. Wise women you are! I love having you around me.

And my best mates, 'The Them', may we always have a totally inappropriate WhatsApp group. Love you all so much – our shared tales of perimenopause woes remind us we have never been, nor would we want to be, 'normal.' Thank God we all know that beige is boring!